Mountains and Minerals/ Rivers and Rocks

Mountains and Minerals/ Rivers and Rocks
A Geologist's Notes from the Field

M. Dane Picard

SPRINGER-SCIENCE+BUSINESS MEDIA, B.V.

ISBN 978-1-4684-6446-7   ISBN 978-1-4684-6444-3 (eBook)
DOI 10.1007/978-1-4684-6444-3

© 1993 Springer Science+Business Media Dordrecht
Originally published by Chapman & Hall, Inc. in 1993

Drawings by Georgia Svolos

Library of Congress Cataloging in Publication Data

Picard, M. Dane, 1927–
    Mountains and minerals, rivers and rocks : a geologist's notes from the field /
by M. Dane Picard.
        p.   cm.
    Includes bibliographical references.

    1. Geology. I. title.
QE35.P48   1992
550—dc20   92-37492
                                                              CIP

For my children,
Marion, Jacqueline, Dane, and Bennet, again;
My father, Vincent Hayes;
And my mother, Velma Vestal Stubblefield Picard

Contents

Acknowledgments

"Tracemakers" and "A Mine Dies" originally appeared in
Earth Science, which has ceased publication. All of the other
essays first appeared in the *Journal of Geological Education*. I
wish to thank James H. Shea and Frances A. Menden at the
Journal of Geological Education; I also wish to thank Wendell
Cochran and Sharon Marsh, former editors of *Earth Science*.
Jeanette Stubbe did the word processing on some of the original
pieces. Craig Sanders has read and helped me with suggestions
on at least half of the essays I have written in the last decade;
his support has been that of a friend and perceptive editor. To
Patricia Faye Cowley I owe more than I can say. She has read,
commented on, and edited almost all of my pieces, many of them
more than once.

The Oil Patch

On the Mountain

As a huge stone is sometimes seen to lie
Couched on the bald top of an eminence;
Wonder to all who do the same espy,
By what means it could thither come, and whence;
So that it seems a thing endured with sense:
Like a sea-beast crawled forth, that on a shelf
Of rock or sand reposeth, there to sun itself . . .

— William Wordsworth, 1807

From the time I was about 4, the words "on the mountain" always had a special meaning for me.

My home then was a sheepwagon or a bunkhouse between West Bridger Creek and Bridger Creek or, for several years, near the East Fork of Dry Creek on Copper Mountain in north-central Wyoming. My mother taught at a one-room school — grades one through eight — and my father ran a small bunch of cattle that the bank began to covet when the winters became harsh. From my mother's schoolroom, near a spring flowing out of the Crow Mountain Sandstone, it was 21 miles to Lysite, the nearest town. When I was old enough to notice, perhaps at 4, it was a great treat to drive to Lysite for groceries.

There were a lot of exposed rocks at all the places we lived; there were also dens of rattlesnakes, as well as badgers, deer, bobcats, and many other animals. If it had rained oftener, there would have been less rock exposed, less attention to the livestock from the banker, and more grass on the range. In the mountains and along the streams, rocks and rattlesnakes were what I noticed most in my childhood.

My mother owned a .22-caliber pistol with a taped, broken, wooden handle. My father gave it to her to shoot rattlesnakes, but she never shot one. She broke the handle clubbing at a snake under a slab of red siltstone that served as the steps to the schoolhouse door. My father said she attacked the snake — which escaped — because I was sitting there.

My mother, in those early years, constantly threw herself between me and the land and animals. It seemed to her I needed protection. I was small, I was sickly, and I barely survived typhoid fever and two bouts with pneumonia. Doc Jewel, the nearest doctor, lived 50 miles away in Shoshoni. In my eyes his most notable achievements were saving my dog Laddie from a rattlesnake bite and fathering Isabel Jewel, a minor Hollywood actress and one of the attendants to Scarlett O'Hara in *Gone with the Wind.*

Through the 1930s there was little prosperity on the mountain or in Worland, the small town we moved to when I started school. My father lost a herd of Mexican steers in a bad winter that decade. He lost his ranch, worked for the WPA (Works Progress Administration), hired out on a farm for 30 dollars a month and lunch, and, in the late 1930s and early 1940s, was deputy sheriff of Washakie County. He scrambled through the Great Depression, a long decade for us, earning enough money to buy cornmeal, bacon rinds, and sacks of beans.

The one-room schoolhouse rested on Triassic sandstone. The rocks of this sequence — named Red Peak, Alcova, Crow Mountain, and Popo Agie formations — were part of the spectacular Wyoming red bed suite. They held few fossils. So, as a boy I did not collect dinosaur bones. Instead, I gathered platey red siltstone rippled-marked by Triassic tides. Then I piled the beds, one on the other into tall sheepherder monuments on the cliff edge east of the schoolhouse.

I searched the area for caves — likely hangouts for rattlesnakes, Mom said. When I finally found a sandstone-walled opening along a fracture, I invested it with a history its size could not support — robber's roost, sheriff's lookout, Indian's signal ground. The small chamber, eroded into Crow Mountain rock by flash floods, was cramped if I carried anything into it besides the gunny sack I dragged along the ground through the sagebrush.

My mother taught me to read before I was old enough to start school. With that head start on the illiterate world I might well have earned higher grades than C's and D's in reading in the first and second grades. I didn't and managed only to earn D+ in Application those years. For a long time, I wasn't ready for school and school wasn't ready for me.

At bedtime my mother read to me and, later, sitting on the edge of the narrow metal bed she read to my brother and me. We snuggled close together to hear all of the story and to keep warm. The Raggedy Ann stories were our earliest favorites; *David Copperfield* became our favorite a few years later.

Grandmother Picard, my father, and my father's three brothers were storytellers. They told stories — understated and laconic — of Grandfather Picard on horseback facing down the sheriff of Casper; and of Harvey ("Kid Curry") Logan, the murderer, riding in and staying a few nights at the 2-B Ranch. In these stories there was little difference between good guys and bad guys. Butch Cassidy, for a time a ranch hand on a spread near Lander where grandma taught school, was remembered as a quiet young man, unremarkable among the other cowboys. Tom ("Peep") O'Day, a lookout for the Wild Bunch, ran out of the bank at Sundance, rushed into an outhouse occupied by a skunk, and was sprayed. The Family Story, told in parts over many years, was about grandpa getting together a large spread and sending the boys and their horses to winter in Pasadena; about the loss of the ranch to bankers; about the later discovery of oil and gas on the former winter range; and about grandpa, in his last years, slowly getting together another stake that he had hidden in a coffee can behind the barn on the 2-B Ranch. It was never full.

My mother's maiden name was Velma Vestal Stubblefield. She and her older brother were raised on a farm in southern Missouri near the Ozarks. Her mother, interested in joining the DAR, had traced the family's ancestry back far enough to get in and, in the process, had discovered an Indian ancestor.

Days on the farm were long and hard and backbreaking. My mother milked 14 cows. She cleaned out the barn. She worked in the fields and in the kitchen. From childhood, she rebelled against the Southern Baptist religion and her mother's domination. Both were tough stubborn women, and their love for each other flared into anger.

My mother completed grade school in 4 years and then took the normal 4 years to finish high school. When she was 14 she earned a scholarship to Southwest Missouri State Teachers College in Springfield, Missouri. With that scholarship she began studies in English.

I have several pictures of her as a student in Springfield. Her hair was thick and short, the color of mahogany. She had full eyebrows and deep-set blue eyes, a slender nose, small flat ears, thin lips, and a square jaw. She was small-boned, about five foot four, never weighing more than 105 pounds. But she was neither timid nor weak. In the 100 or so pictures I have of her, she is always smiling, a smile that is gentle and sad and constrained.

She married at 17, I believe. I was born on August 7, 1927, with hepatitis. She always insisted I saved her life by taking on the morbid condition she would have had, the bile pigments of my blood giving me yellow skin and eyes the color of urine.

She believed we had been abandoned and were on our own. A few weeks after I was born she and I left Missouri for Lysite, then a small settlement of log cabins and frame houses on Badwater Creek in central Wyoming. From there she went to the one-room schoolhouse on Bridger Creek where she became teacher and principal.

She divorced Ernest Hinsley Van Hook, my natural father, a man I never saw. On April 23, 1928, she married Vincent Hayes Picard of Lysite and Holt either in Denver or Boulder, Colorado. Newspaper clippings differ on particulars. One paper carried this announcement: "Mr. Picard is well known in this vicinity where he has grown to manhood, and while we have not the honor of the acquaintance of the bride, we extend to them the felicitations of the community and wish them all the good things in life."

The bride and groom moved to the Goedicke Homestead (where I ate a lot of the calcium-rich chinking) and then moved on to the Kellogg Place on Copper Mountain, about 7 miles on a straight line southwest from Bridger Creek. The elevation is about 6000 feet, near the axis of the range. It is an area noted for snow, long winters, and Archean rock. Cattle do well there in the summer, but there are often only a few weeks of summer, and my father hauled feed to them in the winter. They would have starved otherwise

Before the bank took over the Kellogg Place, my mother sent photographs of it to her mother, with handwritten notes on the back. One of the earliest reads: "My baby, my husband, and my 'Scandal' horse. Also a glimpse of God's country."

God's country is pink and red granite, highly fractured, faulted, lightly mineralized, and older than 2500 million years. The rock is coarse and contains about equal amounts of quartz, microcline, and plagioclase, with 3% biotite, the black mica. Uranium is concentrated in faults near the Kellogg Place in deposits formed millions of years after the granite crystallized. A quarter of a century after my mother moved to God's country the uranium was discovered. By then I was a professional geologist. My uncles asked me every few months for 3 years why I hadn't known there was uranium in the mountain.

Until I was 14, we divided our time between Worland and several places on the mountain. To go to the mountain was to leave civilized Worland for a world of stones and streams and animals. In the autumn it was painful to go back to Worland.

On the mountain my brother and I were seldom disciplined. Away from the house and my mother and grandmother, we ran wild. Our horses, Squealer and Romer, gave us great mobility and we used it to explore the country. The work was romantic — roundups, brandings, and searching for strays; caring parents, grandparents, uncles; adventure: rattlesnakes striking and breeding, bobcats treed by Laddie, an angry bull charging out of the willows on West Bridger Creek, and Squealer shying at a snake and dumping me on green claystone.

I smoked the occasional cigarette, most of them butts thrown out of the bunkhouse or off the porch. I stole a bottle of beer now and then from the spring and felt drunk before I was half done. Grandfather Picard began his breakfast — invariably pancakes, bacon, oatmeal, and black coffee — with a shot of whiskey. No one ate anything until grandpa had downed his shot, set by his plate like a small glass of orange juice.

Mother and my Aunt Evie, both excellent cooks, fixed the big breakfasts, starting before sunrise. We frequently rode all day, coming back for a big dinner. I soon learned to eat enough at breakfast to last the day. Lemon meringue or sour cream pie was what I looked forward to all day on the roundups, and I would trade or promise my brother — a tough bargainer — almost anything for part of his.

When we lived on West Bridger Creek we sometimes rode to an abandoned copper mine at De Pass, on the northeast flank of Copper Mountain. Dad said the copper had formed at an extremely high temperature. That seat-of-the-pants theory — his alone — seems to have come from his observations that the ore was unusually dark.

The mine had been left in sorry shape. A vertical shaft was uncovered; and in the winter, the ranchers lost cows or horses in it. Nothing had been cleaned up. Apparently, the operators had left in the night. The miners had torn ferociously and greedily into the mountain and left it mortally wounded when they abandoned the workings.

Standing on Mame's Hill, we often saw bald eagles soaring. The hill was an island of Eocene rock (Wagon Bed Formation) rising to 6200 feet above a sea of Triassic Crow Mountain Sandstone at 5800 feet. The eagles caught small trout from the streams, and built huge nests on the north flank of Mame's Hill below the tallest sheepherder monument in the region.

Not a compulsive rock and fossil collector in those days, I preferred looking at stones, the streams in flood, snowstorms, wind rippling water, landslides, plants, and animals. My folks, especially my grandmother, had one explanation or another for most objects and processes, explanations arrived at through careful, lifelong attention to the land. The gypsum in the red beds led to diarrhea in the cows. Belemnites, the cigar-shaped fossils in the yellowish and greenish gray claystone of the Jurassic Sundance Formation, were the remains of marine animals; and the sea long ago had covered Wyoming many times, said my grandmother. I kept a few belemnites, and sometimes stuck one of them in my mouth imitating my Uncle Raymond.

Small earthquakes shook the land from time to time. The Wagon Bed Formation, on which the corrals rested north of West Bridger Creek, is mostly volcanic airfall material that buried, or nearly buried, the mountains during Eocene volcanism. Northwest of the bunkhouse I often climbed to the top of a small hill of the volcaniclastic rock. Rattlesnakes summered there, wearing the colorful attire of rocks and sagebrush. The rock colors were in marked contrast to the Triassic redbeds — pale-green, yellow-green, gray-green, orange, pink. Their volcanic origin was not part of the regional folk wisdom and theory.

One of the compensations of middle age is to look back at the raveled threads of one's unique history winding forward through decades.

Fifty-three years have passed since I stayed on the mountain for months at a time. My mother has been dead for 30 of those years. My father died 8 years ago. The rocks of my geological research are the ones I walked over as a boy.

It all seems but a moment.

Looking for Work

There is no worry about running out of oil and gas. The only worry now should be running out of well-trained young men and women.

— John M. Parker, 1982

Don't be pushy or appear to be begging for a job but quietly and gently let the various people you meet at the meetings know you are looking.

— Fred A. Dix, Jr., 1983

The other day a coaster broke off of my chair. I was looking for a piece of paper that would fold to the ½" size to replace the coaster when I noticed a paragraph in the *AAPG Explorer* of an interview with William J. Sherry, Sr. Said Sherry: "The Depression was good for geologists. It taught them realism. It seems like the fellows that start during one of these times do better. They're more daring. Most of the fellows who've been through it think it was a darned good experience."

I say: "Geologists do not walk around looking for a good depression to grow in." But a coasterless geologist on his knees pushing a paper wad under a chair leg lacks authority.

Later on, I considered posting Sherry's interview, but thought better of it. No geologists were hired from among our students last year. This year there have been three. Sherry's words would not have remained the same for long on our bulletin board.

My own experiences with depressions and job hunting in geology began in 1950 except they, perhaps rightly, called that one a recession. In looking back I see the circumstances were less complex than now. I was among a crowd of fresh B.S.-degree holders at the University of Wyoming, one of many geologists and other graduates looking for a job. The class of '50 was the largest class of geologists and other students in the school's history, but it was also the largest class to be unemployed at the time of graduation. No one I knew had a job. A few planned on graduate school — my application had been rejected.

We had talked to campus interviewers, telephoned possible employers, and sent résumés as far east as Tulsa. Most companies ignored the calls and letters. Those who didn't said, "The recession, you know." As if we didn't know.

I had some skill in writing a résumé, which I liked to call QUALIFICATIONS. There was little of my experience I did not manage to fit in somewhere — seismic-crew jug hustler, rodman, surveyor, mechanic's assistant, geologist's field assistant. The companies I had worked for were large ones with appropriately grand names — World Geophysical, Pure Oil, Holly Sugar, and the U.S. Soil Conservation Service. But the pay was small. As a rodman and part-time instrument man for Holly Sugar, I was paid $.65 an hour. The Soil Conservation Service raised this 50%. I considered these wages eminently fair. Job-hunting was a gamble for one who shunned restaurant and grocery store work. I thought an outdoor job was the only real job.

All of the entries under *Experience* in my QUALIFICATIONS were double-spaced. I made frequent changes and frequently rewrote job descriptions. I listed periods of employment to the day. I gave my hobbies — tennis, travel, music. I stretched my height to 6 feet, rated my health as excellent, and noted that prospective employers were looking at the résumé of a man who had served 14 months in the U.S. Navy — seaman second class. Importantly, I grouped these gems under *Personal Data*. I tinkered with different type styles, italics, and spacing. My experience and education were stretched and polished, on the conviction that I would not find work in geology in a recession without a professional-looking and grandiose résumé.

In March of 1950, the Civil Service Commission sent me a Report of Rating. The Expert Examiners — their term — had examined me in December. For grade GS-5, I scored 68 on the written test. My average rating, including veteran preference credit, was 73. No job offer. It seemed unlikely the government would offer a job to a 73, veteran preference credit or not.

In April or May, I heard from the last company. Fine record, they wrote, and they were pleased to have interviewed me, but no permanent or summer job — the recession. I expected the letter, but their tardiness in writing let me rise each morning with a delicious feeling of expectation and powerful fantasies of success in the oil business.

Commencement day, the 5th of June, came with no sign of work. I had no prospects. Like E.B. White, I believed in giving Luck frequent workouts. The local paper reported I would receive the "bechalor" of science from the College of Liberal Arts and so I did. I was waiting for something to happen.

I packed a battered footlocker, stacked books in the back seat of a pre-war green Chevrolet, and headed with my dad for Worland in northwestern Wyoming. At about Wheatland I told him geology is a fine subject to major in, even if I never work a day as a geologist. That was one of those foolish statements I felt compelled to make in those days. I wanted desperately to find a job in geology and I knew very well my dad didn't think any education was worth a nickel if it didn't get you a job.

Two days went by. The next morning Professor D. L. Blackstone, structural geologist at the University of Wyoming, called and asked if I was interested in a summer job doing field work. Texaco needed someone in Casper who could run surveying instruments. Blackstone would recommend me to them if I wanted him to.

Western Union called at 3:32 P.M. John E. Blixt offered $200 a month. I accepted — a dizzying experience, like being down 6-2, 5-1 and pulling a match out 7-5 in the third set.

As weeks passed, I became acutely aware that the Texaco job would end soon — a Wyoming summer is shorter than most. The job was my connection to geology and experience. Following their sallies on the recession, employers liked to point to my lack of experience.

I had acquired a taste for field geology. I wanted to stay with Texaco or go on to another job. I had begun to look at used cars.

The summer's job stretched to the end of November. Blixt, and later E.E. Beeman, like Zeus, decided to prolong the summer and the field season. We mapped through fall days, our fingers often cold. The winds west of Casper brought a smattering of snowflakes.

Beeman tried to hire me permanently. Deane Kilbourne, my field partner, and another geologist I had worked with, A.A. McGregor, extolled my accomplishments in the field well beyond their worth. Beeman sent a recommendation to Denver to Blixt, by then the Division Manager. Blixt endorsed the recommendation and sent it higher, how much higher or where I never found out. Beeman told me the letters somehow reached the Chief Geologist, who decided there could be no exceptions to the policy of not hiring anyone who had not earned an M.S. I thanked Beeman and the others. Beeman gave me 2 weeks with pay to look for another job.

The next morning I rose at 6:00, added 6 months to my résumé, ate, and began to look for work. I soon found others without jobs. The recession was still on. The 6 months of experience added less to my résumé than I hoped.

During the summer the United States Geological Survey had offered me a job in Alabama in ground water. I cavalierly turned it down, telling my friends at Texaco of the disadvantages of life in Alabama, a state I had never visited and knew little about. I didn't care, I often said, to work east of Laramie or west of Salt Lake City. I would move north–south with good weather. I considered myself a Rocky Mountain geologist. There was little point in moving elsewhere until the work was done in the Rockies. Clearly, enough rocks and problems existed to last into my middle years, as distant as they seemed.

An eminent exploration geologist, later a recipient of the Monroe I. Shackleford Medal, the most prestigious award of the American Association of Petroleum Geologists, suggested I take a job in a grocery store near the offices of oil companies. There I could stay close to the action, he said. I should attend the weekly meetings of the Wyoming Geological Association, offering myself for whatever drudgeries arose in their activities. The lectures at meetings were always enlightening, he said. I would make useful contacts. He had worked in a small grocery store in Oklahoma, sorting fruit and vegetables and doubling as a bagger before he got work as a geologist.

The advice was sound. After getting the Texaco job, I had done what I could in the local geological society. Drudgery was everywhere; they were publishing a guidebook. I folded maps, proofread articles, sorted commas for the editor, and carried books from printer to society. To that point I hadn't made a useful contact. I washed off a lot of printer's ink.

My metamorphosis from confident exhilarated geologist to insecure, depressed job seeker was rapid. The first few interviews were enough. Immediate observations: I lacked an M.S. and useful experience. The winter was against me. The recession might lessen in the spring. One never knows when these cycles begin or end, what causes them, or what the companies will do, said a geologist dabbling in predictive economics.

I often sat in an outer office for hours without crossing into the inner sanctum. I often filled out applications without getting an interview. "We'll call you — now hear," became a chorus.

My presence at interviews set off the philosopher in interviewers. A favorite apothegm: "Well-educated and well-motivated geologists will always be in demand." I found myself short on education, long on motivation, and in a field of low demand.

Another bromide, still popular my students say, was the assertion that those who cannot find jobs worry about employment the most. When I first heard this, delivered sagely and apparently without irony, I was near the unsuccessful end of my thirty-fifth interview. Dazed, I went directly to the Henning Hotel Bar and ordered a glass of apple juice.

My strongest memory of the time is of the contrast between warm offices and cold streets. Casper is renowned, even in Wyoming, for the strength and utter variability of the wind and its coldness from November through March. I rushed from the cold into offices, warmed myself, and then went back into the streets.

The 2 weeks of grace with Texaco ended quickly. I had no leads, though I followed every rumor.

Days and weeks passed. No offers of work.

LaVerne D. Hunter, a former classmate at Wyoming, called one day and said there was a job at the Shell Oil Company. They were looking for someone who could use transit and alidade to do field studies on the Navajo Indian Reservation. If offered the job, I could stay within my self-imposed geographic boundaries, pushing off first for Grand Junction, then to Kayenta in northern Arizona.

Success came on my forty-third interview. Alex Clarke, the Shell Manager, said H.D. Thomas at the University of Wyoming felt I could tell bottom from top in most strata. He offered $310 a month and asked if I could come to work on Monday.

To recall those days is like recalling the joy and pain of high-school basketball, of landing and working a paper route with the old *Northern Wyoming Daily News*, of falling in love. I had barely come through. The view was dazzling. You could see the Rockies stretching north to Canada and south to Mexico. The view is less all around these days.

Wind River and Riverton

Humanity is not, as was once thought, the end for which all things were formed; it is but a slight and feeble thing, perhaps an episodic one, in the vast stretch of the universe. But for man, man is the center of interest and the measure of importance.

— John Dewey

That which we do is what we are. That which we remember is, more often than not, that which we would like to have been; or that which we hope to be.

— Ralph Ellison

It was the last third of a long field season. If anyone had said to me it would be the last summer I would run the surveying instruments, I wouldn't have believed them. I assumed geologists would use plane tables forever. But the summer of '50 was the last time I solved a three-point problem. I put away the plane table and alidade for good. An era in petroleum exploration had ended in the Rocky Mountains.

Last days are so emotion-laden that everyone who remembers them remembers them differently. I was 22, a temporary employee of Texaco, hired to run survey instruments on geological field parties in Wyoming. It had been only a few summers since I played baseball and tennis through the summer. I saw myself as a romantic figure, out tramping over weathered badlands while less adventurous young men became lawyers or accountants. I had worked out-of-doors most of my life — paper boy, milkroute boy, cowboy, sheepherder, seismic-crew jug hustler, rodman, instrument man. I was a born field man, a nomad.

We had ventured into the Wind River, the last oil-rich Wyoming basin. There were two of us. My field partner and boss, Deane Kilbourne, was a veteran Texaco geologist who, during the summer, guided me through initiation rites — transforming me from university student to professional geologist.

Everywhere the pastel Wind River badlands were still, even along the steep, shadowed banks of the meandering stream where normally our least movement would have aroused all the inhabitants: cottontails, prairie dogs, and ground squirrels. The mammals hopped across the abandoned point bars, scurried under sparse brush to holes, and skittered wildly on reaches dry since April. Great cracks crossed the floodplain clay — a deserted terrace lightly decorated with flattened stems of late spring flowers, small gray sagebrush, and brown, armored mud balls that had rolled up in the last flood. The sparrows that had sung every morning while we mapped seemed to have left the low dissected plateau or lost their song. The sky was as empty of birds as the sky above George Orwell in the hills round Zaragoza.

Dust lay over brush and rock, a cream cloak covering the foreground indiscriminately. The clay needed rain, but its scent still remained. The dust from our driving had almost settled when Kilbourne lifted his canteen and took his ten o'clock drink 2 hours early.

It was the hottest morning of September. Clearly, we would have to hunt for water, not oil, if we continued to drink through the day at this rate. Kilbourne claimed that some geologists at Michigan State, his old school, would take a large drink at breakfast and work through the rest of the day waterless. That was the kind of story senior geologists told junior geologists who have noticed that Wyoming summers are dry and hot in the basins.

In Kilbourne I had a sensible field partner of calm temper. "Darn it," was his strongest exclamation. He was alert and steady, and he lived his life according to a rigid inner schedule. Didn't smoke. Seldom drank. He carefully combed long strands of his brown hair from one side to the other to make a sparse cover for a wide smooth scalp.

During the 4 days of travel and work since we left Casper, Kilbourne had said little, fewer than a hundred words by my estimation. Eighteen oil and gas fields had been found in the Wind River Basin, all on anticlines and faulted anticlines. We were armed with plane table and alidade the same as the pioneering geologists who found the fields. We knew that hardly any of the prominent surface anticlines were barren. All but four of the upfolds produced petroleum from two or more horizons. Productive reservoir beds were stacked like pancakes and folded into surface anticlines, making graceful arches of sedimentary layers that were a field mapper's dream. We did not know how long it had taken for one of the anticlines to form. We did not know how many undiscovered anticlines there might be. We hoped to find at least one.

In Casper we decided to look for anticlines in the Sand Draw area where the Tensleep Sandstone — a Pennsylvanian deposit in the Big Sand Draw field — had produced about 30 million barrels of oil. Big Sand Draw was discovered in 1918, and 31 years later and 3 miles to the southeast, South Sand Draw. We knew of no North Sand Draw.

Drilling at Big Sand Draw had outlined a large closed area. The anticline was one of a series of northwest-trending folds that plunge into the Wind River Basin from the Sweetwater arch. These asymmetric anticlines have steep, overturned or faulted west limbs. Big Sand Draw was not deeply eroded.

The first few days we looked at the early Tertiary Wind River Formation near Sand Draw. Then, we worked south and east of the draw, climbing a trail and gaining, slowly, the top of Beaver Divide. There and near Sand Draw and Riverton the early Tertiary strata were intermontane basin deposits much like beds we had mapped in the Powder River and Big Horn basins. Our mapping horizons in the Wind River Formation were chiefly yellow-gray and yellow-orange sandstone, pebble conglomerate, and red-banded mudstone. There was a little coal and a few carbonaceous claystone layers. The basin floor had been about 1000 feet above sea level when streams, swamps, and floodplains received sediment from an ancient Wind River range and the Sweetwater uplift to the south.

Red bands were the characteristic feature of Wind River layers. The Fort Union Formation below was gray, yellow-gray, olive-gray, brown-gray, and brown with coal in the upper part. The strata there — dull gray rock filled with live rattlesnakes in sandstone caves set on bedding planes buried in dead leaves and branches — did not arouse the poet in us.

Layers above the Wind River beds were green-yellow, green, olive, yellow-gray — in the sometime palette of a Matisse or Cézanne working in the Eocene with earth pigments. Minerals in the volcanic debris had added their color to the rock and left the signature of extrusive centers that were scores of miles distant and 40 million years before our time.

The principal oil-producing layers in other parts of the basin — some 2 miles below our feet — were the Tensleep Sandstone and the Phosphoria Dolomite. We thought we could find an anticline if an impression of the deep structure was reflected, even faintly, in the surface layers. Few geologists in the Casper Texaco office believed any of the deep structure was reflected in the early Tertiary rock. The largest folds, they said, were bound to have been diminished and obscured by the blanketing effect of subsequent layers and by the earlier beveling during erosion. We mapped because we didn't agree and because our Casper boss, E.E. Beeman, had some confidence in our field approach. He had found an anticline in the Big Horn Basin that was faintly but unmistakeably expressed in the gently dipping Tertiary badlands.

The still land was all legend — animal tribes, Indians, sheep and cattle men, rustlers, the West. We saw scarcely a trace of life north of the Wind River Mountains, along Beaver Divide or near Sand Draw.

In the oil fields they flared gas, and, at night, after a long field day, these lights were melancholy bonfires that warmed neither of us. Perhaps they summoned spirits onto the empty plateau and down into badland draws. To keep their flame alive, the flaring wells burned nightly the energy produced in hundreds of thousands of years. How much gas was lost in a night no one knew (or would say), but it was colossal. These beacons, aflame on every continent, burned more gas than the wells produced.

While we drove along the dirt roads, looking for outcrops, I tried to spot characteristic layers and carry them in my mind from place to place. In other basins it had taken weeks to recognize the sequence of strata. Here, the red-banded Wind River beds stood out, and the layers succeeded one another in a definite order that was soon familiar.

For the Wind River mapping we made Riverton our head-
quarters. I remembered Riverton well from high-school adventures. I played baseball and basketball there, and once ran a slow
mile in a track meet. My dad, one of several Wyoming State
Brand Inspectors, worked the Worland and Riverton livestock
sales rings. I went to Riverton frequently and met a brown-eyed,
black-haired girl who introduced me to post office, a youthful
diversion popular in Wyoming the fall and winter of 1943. I was
bashful and backward. I was a lot more comfortable playing
second base than post office. The young lady sensed this and
gently led me through a furious session of air mail stamps and
special delivery letters. She gave me confidence. I later made up
a poem and wrote her labored letters through the winter and
spring until I met a tiny, blue-eyed blonde from Thermopolis at
a rodeo street dance.

Kilbourne found the best hotel that 5 dollars a night could
buy in Riverton. I found myself on the top floor — the third, if
memory serves — a few feet from the ceiling. I could barely stand
on the west side of the room. The dry hot air collected there
through the day, and by nine o'clock at night the room was a
hot cell, little relieved by an evening breeze. There was no out-
ward circulation at all. The only elements that consistently entered were silt and hot air. In a few days, a loess deposit began
to form on the floor at the foot of my bed. I told the desk clerk
I was a geologist and asked him not to disturb the tiny barchans
that were advancing eastward toward the door, the stairway, and
the dirt street.

I never respected Riverton as much as I did Cody and Powell,
or even Thermopolis. I came to Riverton full of high-school
memories and victories. In basketball we beat Riverton 48 to 27
and 30 to 14 my senior year. In baseball the margin was 10 or
15 runs a game. The *Northern Wyoming Daily News* reported
those wins as "trompings."

Once while I was at bat the Riverton catcher tried to pick a
runner off first base and placed his throw low, just above my left
ear. I saw bright lights and a rainbow over the pitcher's mound.
I staggered but remained upright, determined not to go down.
Frank Watson, our coach, rushed to home plate to check me for

signs of life. Then he became outraged that there did not seem to be a rule to give me first base, moving the runner along. Watson jawed back and forth with the plate umpire — *Daily News* reporting — until his soft pink cheeks turned red and his white scalp broke out in brown splotches. He retreated to the dugout shade. "Stand in!" the umpire shouted. Our bench got on their catcher, ridiculing his ancestry, his knee-high pants, and repeatedly, his weak arm. "Rag arm!" they hooted. "Candy ass!" they hollered, overjoyed that he hadn't flattened me with his throw. I scratched a grounder through the box, a gift it seemed to me from one of the largest Riverton Wolverines, their pitcher, to the smallest Worland Warrior. Later that night, half of our team felt the goose egg above my ear and marveled at the play. The bad throw elevated me for a few days to a prominence I had not enjoyed that season.

I was casting a thin, small shadow in Worland. It was reassuring to see a town that did not seem to be doing a lot better. On days that my jump shot wouldn't fall or I couldn't get the bat around on a curve, let alone a fast ball, I knew I would be quicker and stronger in Riverton.

A long time has passed since our work in the Wind River Basin. I do not remember the errors I know I must have made every day. I thought my journal was for hits and runs, not errors.

Kilbourne asked me to do more and more of the mapping to give me experience and confidence. By eye I traced rock layers, and on foot I tested the extent of those ancient surfaces by walking them. When I dug a trench to place the contact precisely, the contact often wasn't where I had supposed it to be. I decided to dig the contact out at each point we mapped. Kilbourne did the same.

The inclination of the layers was so slight that a very small error in placing the contact between beds could give a completely erroneous strike and dip direction, off by 180 degrees. Few of the contacts were sharp. There were no knife-edge boundaries between sandstone and claystone. After a day of measurements and an evening of calculations and plotting, the results sometimes gave a picture of beds randomly inclined — a map of Tartarus.

The next day we would start again and dig deeper holes to the bases of beds and re-contour our map.

The first oil field found in Wyoming was in the Wind River Basin at Dallas Dome. Dallas Dome lies on a pronounced northwest line of folding, one that parallels the Wind River Mountains on the northeast. The folding is sinuous. The axis is bent or broken at various places. On the west the trend is partly bound by thrust faults that locally override the syncline separating the fold from the Wind River Mountains.

While we worked out of Riverton I decided to see Dallas Dome. Kilbourne took a Sunday off to wash clothes, write letters, and contour the map. I took the car and headed southwest to Lander, then southeast for 9 miles to the field, a great arch of red siltstone and silty sandstone.

In 1884, 66 years before my foray, and only 25 years after Drake's discovery at Oil Creek near Titusville, northwestern Pennsylvania, Mike Murphy, an ex-gold prospector, and a certain Dr. Graf of Omaha made the discovery at Dallas Dome. They drilled a few feet from an oil seep that Washington Irving had described in 1837 in *The Adventures of Captain Bonneville*. Then and now, oil seeps and asphalt attract oil explorers.

Captain Bonneville had "made search for the 'great tar springs' and for the wonders of the mountains, the medicinal properties which he had heard extravagantly lauded by the trappers. After a toilsome search, he found one at the foot of the sand bluff, a little to the east of the Wind River Mountains where it exuded a small stream of something the color and consistency of tar. The men hastened to collect a quantity of it to use as an ointment for the galled backs of their horses and as a balsam for their own pains and aches."

Seeps often indicate a petroleum province of which we had compiled a map in the Wind River Basin. We noted every place we knew of where oil was present at the surface or where gas bubbled in a rancher's spring.

The seep at Dallas Dome was highly prized as have been oil springs and seeps throughout the world for nearly 5000 years. Asphalt was among the building materials in the oldest ruins at Ur in Mesopotamia in about 3000 B.C. Naphtha is a 4000-year-old Akkadian word. Cuneiform tablets from excavations of ancient Middle East cities record contracts for the oil trade and complain about the short supply. The petroleum historian E.W. Owen in *Trek of the Oil Finders* noted: "The prices approved by King Hammurabi's federal trade commission about 1875 B.C. are officially reported. The bricks in the walls of Jericho and Babylon were cemented with bituminous mortar. Noah's ark and the basket in which the infant Moses floated probably were caulked with asphaltic pitch in accordance with the customs of the region. The ghosts of the ancient oil men who supplied the raw materials could have guided modern geologists to the greatest oil fields in the world."

Murphy and Graf saw the seep, walked across the large anticline, and spudded their well precisely on the crest. The crest at Dallas Dome is hard to miss. It is a sheepherder's anticline, "one even a sheepherder could recognize," as Doc Blackstone told us in structural geology at the University of Wyoming. The Chugwater Group crops out on the fold's crest, and successively younger layers are beautifully exposed on the flanks. I walked back and forth in traverses across the crest and then along the axis to the highest point I could reach. Below me the Little Popo Agie River flowed north off of the Wind Rivers in the direction of Hudson and its junction with the Middle Popo Agie River.

Murphy and Graf found oil at 300 feet in fractured siltstone of the Chugwater Group; Triassic strata deposited on tidal flats in shallow seas some 220 million years earlier. The Chugwater field was small, only a few thousand barrels of oil. During the early 1900s other explorers discovered oil at a few thousand feet in the Phosphoria in dolomite and from the Tensleep in sandstone. More than 5 million barrels of low-gravity, high-sulfur, black oil have been produced from these reservoirs.

I learned later that the oil in the Chugwater siltstone came from the older reservoirs, and the oil in them had its source in the Phosphoria Formation. The oil in the Phosphoria came from marine microorganisms in the ancient sea, and the ancient sea came from the west. When I first tramped across Dallas Dome I knew little about source rocks and oil migration.

The structure at Dallas Dome was a colossal one — more than 1000 feet of closed contours (closure) as drawn on elevations measured at the top of the Alcova Limestone. If the great fold had been filled with oil down to the lowest closing contour, the field would have been gigantic. I had a map that showed a closed area about 3 miles long and more than 1 mile wide. But the great structure stretched northwest along its length for more than 5 miles. The Tensleep reservoir (pore space in white, fine-to-medium-grained, quartz sandstone) was more than 400 feet thick. The former sand sea was not uniformly cemented. There were wells with thick pay intervals and ones with thin pay zones or none.

The mornings in Riverton came slowly. I remember waking long before the sun was up. My room was an inferno of color from J.C. Penney's. Kilbourne was asleep, and the thin moon was caught in the west in gray rolls of clouds sailing north. The air was cool. The breeze rustled the gold curtain but did not disturb the apricot-colored quartz crystals the wind had deposited on the window sill during the night. I lay in bed until my mind was racing, then got up and splashed cold water from a metal basin on my face and dried it off with the rough red towel. The purple linoleum had gained its share of the apricot quartz. Below, the street was shadowed and empty. A gaunt, gun-barrel blue cat made its way from the alley mouth to the street corner, then turned east — hunting. I tried to follow, as I had before, but by the time I got downstairs it was gone. Even a veteran cat watcher and tracker can't find a cat that doesn't wish to be found.

I waited for Kilbourne to get up for breakfast and tried to think of ways to improve my chances for permanent employment. Field men were ignored or forgotten by the office geologists, who were the large majority of the staff. Office geologists were close to the district geologist and were close to the oil fields, recontouring each week the major fields in Wyoming. Each office geologist knew he must guard himself and his own. Field geologists were difficult. They did not accept without question opinions of the staff. They just might find a large field and endanger the political equilibrium in the office. They were renegades, practicing a mysterious part of geology that was about to die in the major companies, if the managers had their way. The majority of the geologists with Texaco had never worked through a full field season, and never would.

By the time we got to Riverton I considered myself a field geologist. I still ran the instruments a lot of the time, but had gained considerable mapping experience.

A few days before I first saw the blue cat, I bought what I hoped would be a lucky ring. I paid $5.00 for it. It was sterling silver and moss-agate with a carved arrowhead on each side of the quartz stone. The moss agate had the pattern of half an anticline in the red and brown bands across its center. In my mind I gave the anticlinal pattern more fully developed flanks than it actually had. I was excited to find a tiny anticline, even in a Lander jewelry store. The arrowheads were also a good sign, perhaps a magical sign for the country, I thought. We had done some mapping on the Wind River Indian Reservation. We planned to do more.

There were a lot of mornings I followed the blue cat or some other stray, down deserted streets in Riverton or Lander. They were up early and so was I. I could make out a little of what they said and was sympathetic to their cries of surprise and grievance. Their wildness and grace of movement along shadowed, small-town streets appealed to a lifelong cat lover. When I was 3 my mother gave me an Angora, the first of many cats in my life.

Men and cats are roamers. The cats followed mice and birds and the smells of cafes and bakeries readying themselves for the day. I followed the cats until the sun rose. Then I followed the braided and meandering stream deposits of 30-million-year old strata. I soon saw that, in detail, every deposit was different. Some resembled each other closely — a point bar like another point bar — but I could see how they differed, the characteristic features of each. And later when we drove that way or stopped at a deposit on some other day, I had a comfortable feeling, like dropping by Schultz's Drugstore in Worland on Saturday after a game to have a caramel malt and a grilled cheese sandwich with my friends.

In those days the wet clay scent of the badlands filled the streets of the small Wind River Basin towns and I was at home everywhere. I had played basketball in all of them — Lander, Riverton,

Shoshoni. With my dad and brother I had gone to rodeos every summer in the Wind River towns.

The towns stood as if they knew they could quickly become carbon-rich spots on the floodplains of the rivers at the ragged edge of the badlands. Each town knew it must gird itself to survive floods, rain and wind, long harsh winters, and poor economic prospects in Wyoming. They had seldom known prosperity well or long.

We did a few days of reconnaissance mapping near Dubois in the red and tan, early Tertiary badlands east of town. Beeman visited us there; a ploy, Kilbourne said, for Beeman and two other Casper masters to spend a weekend in the fall on the high, cool mesas. When they came to Dubois we lost our chances for a quiet Sunday on the badlands or in Casper where I had met a girl.

We made short trips from Riverton to check leads that some-one in the Casper office had seen on aerial photos. On one of these trips we drove to a part of the Wind River Basin that reminded me of young Thoreau's assessment of lower and wetter lands: "this curious world which we inhabit is more wonderful than it is convenient; more beautiful than it is useful . . . more to be admired and enjoyed than used. . . ."

Joseph Wood Krutch said Thoreau's proclamation was a common premise of the old naturalists who were less intent upon useful knowledge than upon discovering the hand of God in His Works. The badlands art in that remote part of the great basin gave me feelings I had seldom felt in church. We found no sign of a prospective oil structure. We saw no sign that any man had lived there in the last 100 years.

We traced foot-by-foot the extent and geometric configuration of a meandering stream deposit laid down in a few hours some 30 million years before. Practical petroleum geologists that we were, we mapped key strata. We followed marker beds well-exposed on the treeless slopes. No anticline. No fault. On the dry bare slopes the day was wistful with the far-off feeling of late evening and loneliness.

The modest Texaco salary ($200 a month) and inadequate expense money ($8 a day) were no problem that summer and fall.

It was the best of field seasons. We spent almost all of a day in the badlands. Kilbourne loaned me a jacket to wear on cold mornings. I had more money than I had had for a long time. When there were no measurements to calculate at night, I read geological articles or novels. I kept a journal to produce some sort of record.

I bought a few books in Riverton and Casper — a paperback copy of Hemingway's *Green Hills of Africa* was one. It cost 25 cents in a drugstore. It bore a pale-green cover picture of Hemingway, the bearded author, that was seven times the height of Kilimanjaro, which was reduced to an ant hill as an inset below and in front of the smiling Hemingway. I noticed in the endorsements that John O'Hara said the book was as good as anything he had read in English.

Before buying it I read a few pages while standing beside a display case full of cold cream and talcum powder. The first few pages on writers and writing drew me on. I was an avid reader of literary gossip and advice on writing. In a mystical section, Hemingway maintained that prose (if one was serious enough and had luck) could be carried into a fourth and fifth dimension. This appealed to a scribbler who was mapping near Riverton in three dimensions and reconstructing the fourth from rare fossils, field relationships, and whatever else he could discover.

I soon saw that Hemingway had gotten himself in too deep penetrating the mysteries of the fifth dimension, too deep, anyway, to instruct a writer of casual verse and didactic observation.

While I read *Green Hills of Africa* I moved Thoreau's *Walden* under my 3-pound *Subsurface Geologic Methods*. I carried *Walden* in the field and wanted to protect it in Riverton from Hemingway, who said "naturalists should all work alone and someone else correlate their findings for them."

On a late October day when the air was wine and the sky was pale blue I awoke with a sore throat and aching lymph glands. The morning sunlight lit the bare oak outside my window, and thin frost moved imperceptibly to the window's edges. I gave myself time enough for all the ice crystals to melt before getting up to go find a doctor.

Your lymph nodes are infected, he said, poking my throat and thumping my chest. You have a red throat, he observed. He said there was a lot of infection in Casper in October — in fact, most of the time. I had waited an hour to hear his diagnosis and observations on public health in Casper. Three minutes were enough for him to classify my illness. I had known as much at 3:00 o'clock that morning.

He prescribed penicillin, rest, and liquids. The combination nurse–receptionist–doctor's wife asked for cash, smiling the knowing smile of a woman who deals with drillers and rough-necks, geologists and surveyors, lawyers and clerks. They were all the same to her. I paid up.

I called Kilbourne at the office, then bought penicillin and went home to bed. By then I had a room of my own at Kilbourne's rooming house.

I felt feverish and was cold for 2 days. I pulled the patchwork quilt over my head. The landlady brought chicken broth and tea and a mustard poultice I hid under the bed. I couldn't eat or breathe. Kilbourne loaned me a copy of *The Disenchanted* by Budd Schulberg. I propped my head up on two pillows and read it until I was through and felt well enough to get up.

The chief geologist of Texaco came to Casper. Beeman intro-duced us in the hallway one day as they hurried out to lunch at the Henning Hotel. He looked at me carefully, as at a seaman second class fresh from boot camp, and said he knew of my subsurface work at Salt Creek. If I were to find a possible ex-tension to that field he would be very interested, he said, but these things were hard to find anywhere and I shouldn't get dis-couraged. Beeman looked perplexed but said nothing and led him out the door. The chief geologist had mistaken me for one of the subsurface geologists in the office.

Beeman had kept us in the field as long as he reasonably could. We continued to go out when the sky was a solid, heavy gray in late November. We mapped west of Salt Creek, out in raw wind on Cretaceous claystone slopes and high in dark pines that stirred uneasily. We looked for folds that might have been overlooked by the pioneering field geologists. It began to snow in the mountains near Casper. It began to snow in Casper. I knew that Texaco could not carry an instrument man through the winter, no matter how ready he might be for a winter camp on the Powder River or on Walden Pond: "When the ponds were firmly frozen, they afforded not only new and shorter routes to many points but new views from their surfaces of the familiar landscape around them."

Kilbourne told Beeman that I was a sounder geologist than anyone else he had worked with in the field. This exaggeration impressed Beeman, and he asked the chief geologist to keep me on. The chief geologist said no and mentioned that I did not have an advanced degree. Unappealable operating policy of Texaco, the chief geologist said. Beeman gave me 2 weeks to look for another job.

My metamorphosis from instrument man to job hunter was painful. We had had exceptional luck in the Wind River Basin. There was a North Sand Draw anticline expressed in the rainbow-colored stream strata. We had tramped up and down a score of canyons and across as many buttes to map it. Its contoured form in gray and white on our land grid increased its importance in my mind. It was no longer an idea or a dream; it was reality. I thought the thin pencil lines might save my job.

Our Casper masters showed little interest in North Sand Draw or in a prospect we mapped near Riverton Dome. They bent low over their desks and light tables to re-contour oil fields that had been productive for decades. Major oil companies look for large prospects. Texaco was hunting mice with a thirty-ought-six.

I tried to exorcise my misery with a verse:

In Wind River badland layers
I look for form
The ghosts of folds
Flexed and frozen
In deformation.

Anticlines are hard to find
That hide under covers
Where limbs are smoothed
Their crests subdued
Below pastel Tertiary strata.

Oil hunters reach deep
To draw contour lines
Of flanks defined by dip and strike
Warily expressed above timeworn
Arches undrilled by anyone.

I think of how
I traced in sand and clay
One fall North Sand Draw
Across rock axes,
Up and down their plunge.

Geologists like to close
Contours and drill
The chaste dome
They know in basins
Their rivals little court.

Here on green stream strata cut
In black mica and scarlet quartz
I listen for rotary sounds:
Drill this structure, Texaco, please.
A wild rock hammer longs to ring.

Youthful verse and wishful thinking. North Sand Draw apparently slid soundlessly and forever into a large, inactive bin in the district office. It became a small oil discovery of another company in 1953. I never knew what happened, if anything, near Riverton Dome. I earned $1400 mapping Wyoming badlands that season and added 7 months under *Experience* in my QUALIFICATIONS — all double-spaced. I was very young.

The Lady with Red Shoes

The most exciting event, or at least the most talked about event in Vernal, Utah, that summer, took place in the Hotel Vernal lobby at 2:00 in the morning. As one of two eyewitnesses, I was called upon many times to tell the story.

The company I worked for as well-site geologist had spudded a well at Red Wash. I was assessing well samples from all over the basin and studying Eocene strata in the field, and had been living at the hotel for several months.

The drill had reached 4900 feet. I received a call from the rig at 1:30. I dressed, splashed cold water on my face, and took the front stairs to the lobby. I put my gear in a blue Chevrolet rental car renowned for its thirst — giving 4 miles to the gallon — and for slow starts, and choking and gasping on the warm June mornings. Once again it was reluctant and, finally, unwilling. I stomped back into the lobby to give it a chance to pull itself together. I stopped to ask the woman at the desk if she thought *FBI Girl* was going to play all month at "The Showhouse." She thought so.

Sometimes in the evening I went to a movie. Sometimes I read *War and Peace*. Watching the well, I was losing my way in *War and Peace*. My mind was confused. When I sat in the coffee shop or walked through the town, I wanted to meet attractive women. There were few chances for an out-of-towner to socialize in Vernal. I had no desire to see *FBI Girl* again.

I strolled across the lobby and picked up the *Salt Lake Tribune*, almost 24 hours old by the time it had made the bus ride to Vernal. I didn't care how old the politics and murders were, but a day for me in those years was too long to wait for baseball scores. The radio in the Chevrolet worked, which gave me a chance to get the scores, if not the statistics, the day of the games. But a Giants fan wants to know everything, whether Willie Mays went 2 for 4, homered, or singled twice. A good newspaper gives enough statistics — the box scores, the standings, the batting and pitching averages on Sunday — to last 20 minutes on weekdays, and an hour on Sunday.

I paid for the paper. The desk clerk asked about the well. In a boom, local people are as curious about what is going on as an oil company scout. They sometimes gather and sell information themselves.

There was little I could tell her. We were drilling the well tight — no information to anybody.

We talked another 10 minutes. She had been working nights for years and had grown accustomed to graveyard shifts. They couldn't get anybody else to take it, she said, but God knows, though the night shift is generally quiet, you were knocked on your butt by sun up, the same as after 8 hours of day shift.

I told her I'd felt great when we spudded the well, but had gotten tired when I worked 3 days with 6 hours sleep, and now I was beat all the time. The phone rang in the night. The engineer pounded on my door. Too many things at the well took place at 3:00 in the morning. I said I didn't think I'd ever feel rested again.

As we talked, the door to the lobby opened and the cool night air rushed in — air and dust from Main Street. I saw her eyes widen. I looked toward the door. A young blonde woman was strutting toward the desk dressed in a pair of new red high heels with ankle straps. She was statuesque. She looked straight ahead. Except for the pumps, she was naked.

Geologists watching wells, at least in those days, had little experience with nude blondes sauntering across hotel lobbies. Yes I was excited. I looked away, then back gain. What does one do or say? Half stunned and yet fascinated, I thought of saying good evening or hi. Should I tip a nonexistent hat, or just stare? I stared and smiled. I tried to keep my eyes on her blonde hair.

She asked for a pack of Camels. She held the change in her hand. I remember thinking I should warn her that Camels might be the worst brand going, loaded, if she did not happen to know it, with nicotine, brown gunk, and early morning coughs. I smoked them myself.

I said nothing. She stared back but for a shorter time than I wished. She paid for the cigarettes, wheeled, and her red high heels click-clacked back across the lobby and out the door. Her back was straight, her shoulders square, and in the bright lobby lights her skin was white as Carraran marble.

We watched her get in a green Ford at the curb and drive east down Main. I began to laugh. The desk clerk laughed. We talked and laughed. She knew the woman's name, and so did I. Nearly every day I saw her driving up and down Main, always clothed, however. I saw her in restaurants and the hotel's private club. I hadn't met her, and never would. But I never quit expecting to see her in the lobby again.

The story of her stroll became the talk of Vernal. Later on, I heard that several drillers had seen her driving along Main before she stopped. A roughneck reported seeing her turn down the golf course road after she left home.

When I remembered to try the car, the carburetor was well. The engine turned over and caught. I drove east toward the Green River.

Her presence filled the car, crowding out thoughts of what might be happening at the rig. Was she somewhere along the road? I had trouble remembering where I was going and why I was driving east and not south. I wondered if she had undressed at home or in the cool night beside the green Ford. I wondered why she had strutted into the lobby and if she had found the motions of spring air on her body exhilarating. I rolled down the front windows to let the night air rush over me.

The nude lady had been the subject of talk in Vernal even before that night. She was the most talked about of the town's most attractive young women. They were in their twenties — my age — and I had met most of them, but knew none of them well enough to ask one to see *FBI Girl*.

Her face began to fade. I became fitful, trying to remember everything. I reached the Green River, crossed it, and turned south onto the dirt road to the rig. I remembered details of the call bringing me to the well. A few sandstone chips were saturated with oil that filled pores, the mudlogger had said.

Weeks of getting up in the middle of the night had made me kind of crazy. Driving from the rig in the evenings, I stared too long at figures carved in sandstone beside the road. At night, I saw giant faces in mist on the river and swerved suddenly to avoid them. Had I really seen a blonde in red shoes, or had I imagined it?

At the well I studied the sandstone. The porosity was low, but oil staining was uniform. Enough to test.

I called in the testers and drove back to the rig. I wanted to tell the mudlogger about the woman, but then decided I wouldn't.

The first light touched the tip of the rig, lighting the derrick. Struck by the sun, everything was beautiful. In the light, I thought I saw a red-slippered lady dancing on a crimson lake. A mud pit had never looked so inviting.

We recovered three barrels of slightly oily, gassy, drilling mud on the test. No discovery. Not even enough oil in the mud to color the pit.

A few hours later, I sat over breakfast in the coffee shop and recalled the previous 8 hours — could it really have been only 8 hours? So much had happened. My first time to test for oil and the first time I'd ever seen a naked lady — in public.

That night I was intensely receptive to everything. Late the next morning, after my first night's sleep in days, I found I could no longer fully summon — try as I would — the Lady Godiva in new red shoes. For the rest of the summer at night I came down the stairs to the hotel lobby in anticipation, and I spent hours looking for green Fords.

Walking the Four Corners

The air was dry and still, but the heat moved in waves. If you reached out you could feel the heat striking your arm. My clothes were already soaked with sweat.

— Eddy L. Harris, 1992

At the outset of my second year as a professional petroleum geologist, my truck broke down and I walked 27 miles across the desert of the Four Corners area, the point where Utah, Colorado, New Mexico, and Arizona meet. All of my walk — thereafter I called it The Long Walk — was in Utah, northwest of Hatch's Trading Post and east of Blanding, a town of several thousand people. I walked through the late evening and night, alone with the rabbits and coyotes and my thoughts.

That was in July 1951. At 23 I was young to be in charge of a field mapping party, even a party of one. I was mapping for the Shell Oil Company, looking for anticlines, the upfolded strata petroleum geologists spend their lives seeking. To be mapping alone in the desert or anywhere 50 miles away from the nearest town was a poor idea.

Ed Wright and Dave Shoemaker, two other field geologists with Shell, lived in a camp on Montezuma Creek, but they had gone south into Hovenweep country. Later on, they planned to meet their wives who were coming for the weekend to Blanding from Grand Junction, the location of Shell's district office.

When the truck wouldn't start after one of my stops, I began to worry that Wright and Shoemaker wouldn't wait to see if I had gotten back to camp. They hadn't seen their wives in 2 weeks;

they had talked of nothing else for a week. The desert offered no respite of the sort they craved. Nobody was coming to my aid.

I tried to start the truck and finally got it to roll a few feet downhill. But that was it. The tires settled comfortably on brown, rough-edged sandstone. Another failure with machines. I never was good with wheels and pulleys. I have been careful to keep my hands out of the innards of such contraptions, which may be the reason I didn't learn to drive a car until I was 20.

My water supply was down to three-quarters of a canteen. I sat in the shade of Jurassic sandstone and opened my lunch sack — half a can of Vienna sausages (brown from sunburn brought on by a two o'clock temperature of 107 °F), an orange, and brown bread, crisp and hot and flavorless. I ate the orange.

The truck quit at two o'clock in the afternoon. I decided to stay in what shade I could find until about six o'clock. In the afternoons there wasn't a lot of shade under the sandstone ledges or moving air in the hot narrow canyons. Every week of mapping seemed hotter than the one before. The glare of the sun was everywhere.

I was tracing contacts between rock units in the Morrison Formation — badlands strata noted in the Rocky Mountains for dinosaur skeletons that occasioned the acrimonious exchanges between Othniel Charles Marsh and Edward Drinker Cope. I was trying to determine the structure of key horizons in the Morrison — whether they were folded into anticlines and synclines — as a clue to the deep structure of older layers in the basin. We thought oil was present in Pennsylvanian rocks, which are about 150 million years older than the Morrison and 5000 feet deeper in the earth. I kept in mind that E.L. Goodridge had discovered oil seeps in Pennsylvanian rocks on the San Juan River near Mexican Hat in 1880, an indication possibly of larger deposits elsewhere.

While I waited for the sun to go away, I looked through my notebook. I had found few anticlines, and none that weren't previously recognized by other geologists. As a note taker I was cryptic, leaving to the imagination, or to memory, important details of the geology. The poorest notebook is better than the best memory; this is what I was taught at the University of Wyoming. Apparently I hadn't learned that lesson well. No manager I worked for ever looked at my notebook, which may have prolonged my career in petroleum geology.

I had seen "yellow sandstone" in the Morrison, north of the Hatch Trading Post. Perhaps these were from iron-oxide-rich layers, perhaps they were carnotite-bearing sandstone samples. Later, prospectors found uranium and vanadium minerals in the area and several mines were opened. I collected a few chunks of the yellow sandstone and let it rattle around in the truck for a month before throwing it out.

While I looked at my notebook, in the afternoon's gentle decline, I drank water. I seemed to be thirstier than usual. I was not a fieldman who could stride through the day waterless.

The braided and meandering streams that had deposited detritus in the Morrison channels flowed once again on my map and sometimes to bankful in my thoughts. Braided streams were dominant during Morrison time. Lakes occurred on the floodplains. Volcanic ash fell in large amounts, particularly late in Morrison history. The aggrading deposits formed on a series of relatively featureless surfaces. The abundance of gravel in all parts of the Morrison suggested mountain-building in the source areas.

I walked southeast out of the dry canyon where I had been working. The heat had subsided hardly at all. Clouds were few. A rabbit moved beneath the overhanging wall of one of the large channel sandstones. A bird in a cottonwood sang close by. The evening world was moving quietly below the sparse brush.

I came down the draw and across pastel badlands of shale and siltstone, a still land now compared with the clammer of sound and life during the dinosaur's reign. The air smelled of clay. The draw had been dry so long that it had broken into polygons along the bottom.

I climbed out of the draw and walked on into the large drainage system of other Montezuma Creek tributaries, also dry and mudcracked on the floodplains above their empty channels. Brilliant light bounced and quivered ahead of me. I saw key beds I had traced in the mapping area but they, too, moved up and down with the bright light, giving a mirage-like motion to surface structures I might have mapped. If an anticline existed there, its relief seemed too slight to withstand the movement of light, its contours forever warped in the thin desert air.

Lizards watched my progress. They knew the secrets of survival. I reminded myself it was going to be just a long hot walk. The low outcrops stretched ahead of me above the river's floodplain. The dark mass of the laccolithic Abajos rose to the north, blue in the distance, and steeply inclined in their highest parts to eight independent summits. The peaks were rounded, some nearly flat. Sides sloped steeply and evenly, and, near the summits, changed into long spurs with broad tops and smooth sides. There were no bold masses of naked rock.

I was a speck of life on the Sage Plain, a plain that extended eastward from the Abajos and Comb Wash to the Dolores River and the Mesa Verde in Colorado. Profiles drawn east–west across the plain departed from the horizontal very little. If I stopped too long to rest, and a sudden desert windstorm arose I might fly eastward like a silt particle lifted across the plain to a Mesa Verde cave. While the sun remained, almost any point afforded an uninterrupted view of distant features: the Mesa Verde wall; Ute Mountain west of Cortez, Colorado; and Shiprock in northwestern New Mexico, a distant galleon plowing heavy seas of gray Mancos shale.

In contrast, a profile drawn north–south from Monticello to the mouth of Montezuma Creek on the San Juan River sloped about 2000 feet in less than 50 miles. I was walking downslope to the south, of which I reminded myself whenever I felt desperately tired.

After the first few miles I was thirsty. I remained thirsty the rest of the way. It was like the thirst of 20 minutes of hard basketball with no time-outs. I hoarded the water in the canteen, taking only one swallow each time I stopped to rest.

I imagined all the streams flowing again: the ancient Jurassic ones, and the latter-day dry washes of Montezuma Creek and its tributaries. The braids would fill slowly with rain swept down from the Abajos; the channels would coalesce as the storm continued; and a sheet of water finally would fill all the scarred channels. I would cross Montezuma Creek, as one might cross the braided San Juan River in places. Standing chest-deep in the water of the rejuvenated streamways, I would drink when I wished, as much as I wished, part of the living river flowing out of nowhere into nothing.

The hours passed slowly. I began to drift, not toward the trading post and water, or to a cool July beside a mountain stream in Wyoming or Spain, but only toward Sunday morning.

The great plain was deeply dissected by stream courses that began as canyons and remained canyons all the way to the San Juan River. Narrow shallow channels in the mapping area were vertical-walled because the initial trenches were cut through thick beds of resistant Morrison sandstone or, higher in the stratigraphic sequence, through the Cretaceous Burro Canyon or Dakota sandstones. Canyons 100 feet to more than 500 feet deep appeared without warning. I could cross only at rare places where the tops of their enclosing walls were broken by cracks or fringed with talus. I could take a wrong turn.

When the sun went down it was cooler. I kept telling myself it was true until finally it was. By then I was out of the short canyons, nearly due north of the Hatch Trading Post, and could walk along dirt trails and roads.

On an empty road at midnight a meteor trailed high above me, dissolved, and vanished. I took it to be a good sign. It reminded me of the flying saucer we had seen that summer. Or so we thought. Flying saucers were a rare sight then, though now they are so commonly seen in California that reports of them receive little notice.

The one we saw was last seen by us at 10 P.M. in June of 1951 flying west at a steep angle of climb, north of Ute Dome in Colorado and about 17 miles west of Cortez. My journal is not much further help with this. I may have been wary of recording all the details. My commitment was to science. But a diarist, down deep, lives with the delicious anticipation that his work will end up bound in buckram, rescued from obscurity for an eager audience in the next century.

The sighting was confirmed by Wright and Shoemaker. In fact they saw the saucer first. I was sleeping in the back seat of their truck. Wright was driving. I was at the beginning of a dream sequence about muskrats, disturbed by their shouting. The saucer, blue lights blazing, was between us and Ute Mountain, climbing steeply in the direction of Mars. My view was a brief one. A swirl of low clouds obscured my sight. The craft was flying at a velocity suggestive of a pilot late for Los Angeles. For the remainder of the evening and for several days afterwards we talked about the details. But they remained fuzzy for me, a sight seen too quickly, too unexpectedly.

As a diarist, I was a master of suspense, leaving to future readers the challenges of filling in whole sequences of my improbable adventures. My facts were bare, unadorned with narrative extravagances, suggestive of my enthusiasm for T.S. Eliot's poetry and the spareness of William Carlos Williams.

The night of my desert walk had a strange beauty. The moon's oblique rays lit up a landscape heavy with shadows and a weary grandeur. The distant peaks of the Abajos, on which the moon shone its brightest light, had become silver, and the few high clouds that hovered above the mountains and the valley floor were silver too. My sense of isolation was heightened. I was conscious of the brilliance on the plain, of being on a limitless plain with the same night's journey always ahead, and of the dark shadows covering a world I could only hear.

When I sat down to rest, sometimes stretching out on a pastel spread of green and violet claystone on the floodplain, the night animals moved closer to inspect. I saw eyes of animals I couldn't recognize. It seemed they were conscious of my fears, my unsuitability for survival for very long in their world.

After two o'clock in the morning my water was gone. I sucked on polished pebbles from the Morrison conglomerates, perhaps on a gastrolith, a stomach stone of a rare juvenile dinosaur, a *Camarasaurus*. The stones were exceedingly smooth, too smooth and polished, I thought, for them to have been so highly abraded in the ancient stream channels. I liked the taste of the gastric-juice polish on the quartz pebbles. But perhaps they were not stomach stones. I do not know. Some geologists call them gastromyths, probably for many reasons.

I had hoarded the water long enough for it to last into the cool night. My legs grew heavy and my head became light. I wanted to drop down and wait for the sun. I wanted to put my feet in a pan of cold water and drink iced tea while reading the box scores to see if Yogi Berra had hit one out of Yankee Stadium. Water in streams, water in pans, water not to be hoarded — water not to be remembered or thought of in the course of a day.

As I trudged on, safely off the obscure trails and on a dirt road beside Montezuma Creek, geology seemed a not very romantic occupation. This was a thought completely contrary to my usual state of mind. I enjoyed walking day after day from outcrop to outcrop and grew browner, leaner, tougher — an explorer, a small figure in a long line of North American explorers. I felt intensely alive, exhilaration sweeping through me when I found an inclined layer, vanishing here, reappearing there, perhaps the folded flank of a large structure not seen before, one invisible to other eyes that had traced rocks in the Four Corners. Geology gave a religious feeling without religious images.

In the cold, bright night, I felt the inadequacy of my clothing, items better suited for mapping in Wyoming. I had the wrong kind of boots and a light nylon jacket that would have looked great on Jimmy Stewart, but was scant protection through the coldest part of early dawn on the desert.

In that light, the stone figures along the stream course began their morning retreat. Soon the shadows held no threatening apparitions. Then a surge of energy propelled me crazily down the road, energy I was unaware of an hour before.

My feeling of elation grew and I walked the last half mile jauntily. My fears departed. I stood a moment looking back the way I had come and saw the land lighten. The Hatch Trading Post was around the next bend.

I walked to the corral and slipped between the posts. A sorrel horse gave a sudden snort and turned away. I knelt by the water tank and stuck my head in. I was soon drunk on water, the faint taste of algae giving the cold water the sweetness of French sauterne. When I was a boy in northern Wyoming I had typhoid fever; the water tank held no other possible perils for me. At last I was no longer thirsty. In the coolness of the morning I walked on to the trading post, the horses staring curiously at the harmless madman disappearing into the morning desert mist.

Academe

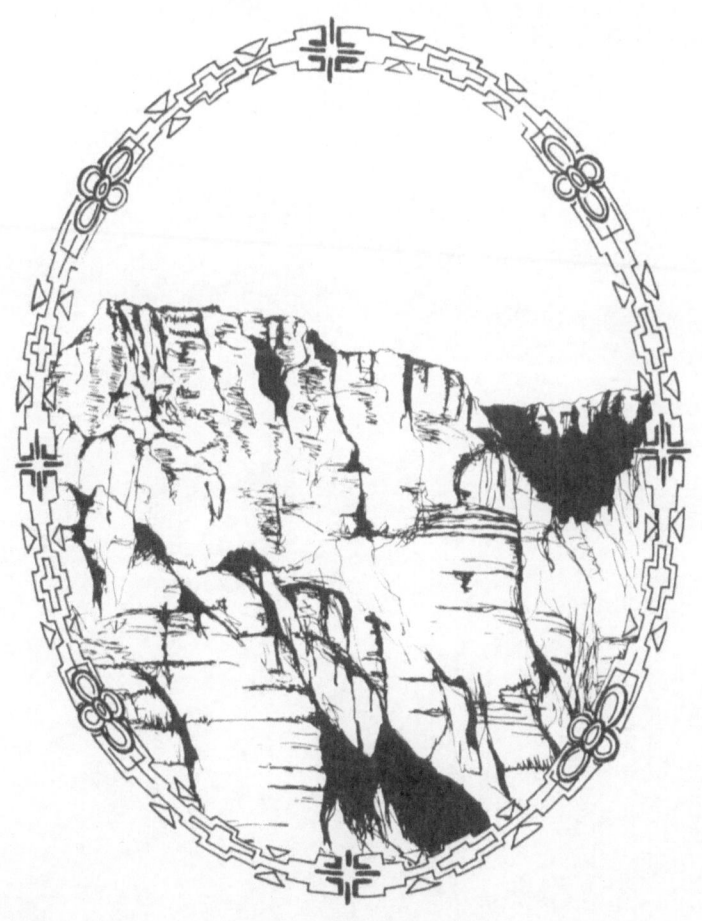

Appointments and Reappointments

The most spirited faculty meeting of the season took place last Tuesday. Such events are not always recorded, for which we may be thankful. In this instance, however, the group agreed that we needed careful notes lest we forget a major decision or a pivotal detail. By acclamation and the chair's suggestion, the job was mine.

No one but me, apparently, remembered about my handwriting. It hasn't been legible since the eighth grade when I won a Palmer Penmanship Award for rolling my wrist in the Palmer style through a timed 10 minutes of the Declaration of Independence without misspelling consanguinity or usurpations. Most days my writing looks like a prescription for an antibiotic written in turbulent flight over West Texas. Frequently, I can't read it myself.

Here then is an unofficial account — much amended — after its passage through the faculty for additions, deletions, emendations, and immersions. One of the junior faculty spilled sugar-free Dr. Pepper or coffee — I'm not sure which — on the first draft. The draft pages were stuck together, seemingly forever. Twenty-five words on the second page were blurred, and the first two paragraphs were blotted in the flood. I managed to recover them by using a light table and looking at them at an angle of 30 degrees in reflected light at midnight. Other procedures might have worked as well.

The meeting did not start off well. There was more than normal confusion about who was in charge, the outgoing chair being on some kind of leave, and the assistant chair having just returned from leave. Finally the assistant chair stepped briskly forward, scattering papers and untenured professors.

We reached a quorum after 20 minutes of waiting. It was 74 degrees outside and 66 degrees inside, a glorious spring day — the kind of day a lot of the faculty go to the library, presumably Berkeley's because no one can ever find them after they leave the building.

For those who had forgotten to bring the memo or who hadn't gotten around to reading it, the Chair (Assistant deleted from here on) said we were meeting to appoint or reappoint — as the case might be — auxiliary faculty. Their names, he said, were listed by various categories — Assistant, Associate, Professor, Adjunct, and Research. The chair reminded us that we could add names to the list if we felt compelled to do so by mustering a second from the floor and gaining approval of the faculty. Everyone nodded — whether in approval or early afternoon drowsiness, I was not sure.

Before we had moved very far into deliberation, Professor Johnson suggested facetiously that we ask the auxiliary faculty — which number about the same as the regular faculty — if they would teach fall quarter while we took the quarter off. This drew a titter. Johnson is about as witty as we get. He is the one who last year proposed that the next faculty opening be filled with a psychiatrist. He also suggested that department heads serve 3 years, then wander out to Mt. Shasta and sit contemplatively on the summit monitoring seismic tremors. That sort of thing.

The first few reappointments went quickly. These were Full Adjunct Research Professors. If there was some difficulty in remembering their research achievements, it didn't cause a lot of concern because each had a champion on the faculty. Some of the faculty could not recall anything several had done and wondered if their appropriate level wasn't closer to Assistant or Associate than Full Professor. No matter. No one really expected them to spend any time in the department. It was too nice an afternoon and too early in the meeting for Johnson's suggestion that we should drop them because of slight professional reputation and negligible contribution to our program.

The chair excused himself and turned the meeting over to Professor John Preston Worth, a former assistant chair and senior member of our group. Worth spread enough papers and memos in front of him to bankrupt a small college's office budget for a year.

The next candidate considered was listed as a Full Adjunct Teaching-Research Professor. His name was known far and wide. Professor Worth murmured that this professor's name would go down in history. He went on in the same vein, mentioning papers published, invited keynote speeches given, grants awarded, conventions attended, and field trips led. Some of the faculty had dozed off when the chair left and Worth began to speak. The rest had learned through the years to tune Worth out.

"Very likely you are on the mark," Albert Bloodright finally said. "Everybody knows of the man's contributions to blastoid research and his stimulating lecture series in the 1950s on very slow evolution." Bloodright looked challengingly around the room, a frequent tactic to control junior faculty, to seek support for his opinions, and perhaps — as Johnson once suggested — to try to find the track again. Half the time Bloodright will tail off after one of his outbursts. This time he didn't. He acknowledged Vincent Van Hook, a young paleontologist, who was awake and had become more and more agitated as Worth and Bloodright spoke.

"Didn't Professor Winslow die 2 years ago?" Van Hook blurted out. It was an embarrassing question. Several of us had wondered when it would come up. I had helped prepare the obituary, which was fitting, if a trifle short of insuring Winslow's immortality much beyond Bountiful.

Worth and Bloodright showed little embarrassment. Worth hedged a bit. He wondered why he hadn't been told. How was it that Winslow happened still to be listed in our catalogue? Bloodright straightforwardly recalled that he was out of the country at the time. Yes, that was the trip to one or the other of the African countries to look for vivianite.

The chair returned; the rest of the faculty awoke. The majority enjoyed their colleagues' discomfort and were sorry to see it end.

Bloodright had stung them all too many times for them to show their charity.

We decided unanimously to reappoint a supposed earthquake specialist from the U.S. Geological Survey. Bloodright said the man was an important contact in that department, although he was uncertain of his current role in our department. "He may be interfacing with the earthquake group and the structural geologists," Bloodright continued, "and they interface with the State and Federal surveys." Johnson asked if we now had an earthquake authority for each television channel. He said there had been considerable grumbling and some panic during the Challis Idaho quake when one of our authorities had ducked out to play racquetball, which had left Channel 49 uninformed and uninterfaced.

"In a big quake the adjunct appointee might be useful," said Van Hook. Everyone agreed with that. But there seemed to be enough authorities in the department for most television and radio appearances most of the time. "In a small quake, we would acquit ourselves splendidly," said Worth. "No department in the Rockies can handle small tremors so well."

We talked for another 10 minutes. The chair reminded us that we had already voted to reappoint the earthquake specialist.

Faustino Franco, one of our seismologists, moved for reappointment of our woman adjunct professor. J. Harrison Snubbers asked if anyone was concerned about her publication record. Franco said she was publishing at the rate of six significant papers a year. No one else was aware of this. There were no files available on the adjunct faculty, and no one had ever asked for them.

Johnson, who surreptitiously and habitually reads Mark Twain during faculty meetings, said "Woman's equality with man has never been conceded by any people, ancient or modern, civilized or savage." Van Hook concurred: "The world was made for man, and the universe was made for the world — to stiddy it you know."

In the uproar, Franco's motion passed and two people slipped out the back door, off for home or a coffee shop.

"It would strengthen our hand to have a geotechnical person," said Professor Sampson. "Comparable universities on both coasts, in the Midwest, on the Continent, and — so I've heard — in Russia and China have close ties to geotechnicians."

For the next 20 minutes there was much discussion of geotechnical people. Three-fourths of the faculty did not know what Sampson meant or wanted. Puddingstone said there were geotechnical people in the Engineering Building, implying that we might capture one after dark with a net or during the day by offering a modest consultancy. Unable to quickly find an apt quote, Johnson improvised: "Illustrious geologists and geotechnicians buy these appointments for 999 years, just as they would buy a building lot, and they pay $250,000 for a Full Adjunct Professorship and will transmit it to their oldest son or only daughter, whichever the case may be."

Sampson sniffed. The junior faculty laughed, at some risk to their respective and forthcoming reviews. We passed a resolution to look for a geotechnical person. We would look first among our alumni, then in engineering, and finally in industry. The chair recited our vow of poverty — no funds for the adjuncts ever, and none for yellow writing tablets until July.

We sprinted along at a fast clip, nearly finishing in 3 hours. At the end, half of the faculty were milling about the room, bumping into each other, pushing chairs out of the way, and trying to open a window sealed to keep cold air and students in the building and noxious fumes and particulate matter out. The chair, a former nose guard on Nebraska's national championship team of 1971, restored a semblance of order by jovially cracking Bloodright a hard one between the shoulder blades and sitting him down. That demonstration of parliamentary power was well taken. Calm and relative quiet descended. We began to sign the appointment forms — pink, yellow, pale blue, white, and a splendid new sepia sheet — which didn't take much more than 45 minutes.

I don't believe I've ever seen the faculty so pleased with their efforts. They left the room slowly, even dreamily, and wandered out onto the campus, some sitting down on grass mounds — our last president's contribution to campus landscaping — and others leaning against gneiss boulders and siltstone slabs that are decoratively strewn about. Many were still there an hour later, reluctant to break the mood established in the meeting. There was a little concern among all when Puddingstone toppled over and rolled down the flank of a mound. But the chair quickly pulled him to his feet, picked up his hat, and led him homeward.

Bloodright said our president had gotten the idea for the landscape engineering from looking at photos of glacial terrains in Alaska. "The mounds resemble drumlins as much as anything else you'll see in Utah," he concluded.

"These are far from being as elongate as many I've seen in northern Manitoba," said Worth, as he and Bloodright made their way through the pseudo-drumlin field toward the library, a modern building of concrete and glass and Italian marble.

Adding a Specialty

My colleague feels he needs to broaden his research, and has decided to add deep-sea coprolites to his specialty of continental coprolites. Though some paleontologists consider them dull, continental coprolites are a great little group.

But most of his world says no. His chairman says he should stay in the field he was hired to represent in the department, but he warns him again that he is expected to bring in research money, write a couple of dozen significant papers a year, be an excellent teacher, be a good citizen, abstain or drink only moderately at student events, and participate in outside activities — geological societies, invited lectures, retreats, parades, Mah-jongg tournaments, and so on. He already works 68 hours a week for less each year than a popular divorce lawyer makes in a month.

Many of his faculty associates wonder if he is not partly withdrawing prematurely from a rich (even fetid) field and, ultimately, from his close associations in the department. His wife fears that there will be emotional and economic repercussions. By adding a new specialty, he may endanger his children's future, his retirement, his family's life style, and the intimacy of his marriage, she says. I remind him of one of the department's horror stories — the sedimentologist who tried to switch from studying structures on the tops of sandstone beds to studying the bottoms. He no longer goes to the field, not daring to be more than 10 minutes away from his psychiatrist. His family has become dysfunctional, his wife affirms, citing a new profile of dysfunctional families in one of the dozen books she carries in a bag large enough to hold the collected works of Jung and Freud.

I am one of the few in the department who still supports him — let's call this romantic paleontologist Fred — in his wish for a change of direction in his research. I support all who long to become geological Renaissance figures, studying a reptile here, a chunk of red shale there. I know perhaps better than most of his colleagues the difficulties Fred is having in trying to join a club that doesn't want any more members.

What Fred needs most is financial support for the new studies. Because his research on continental coprolites is not as valued in the department as, say, the work of the team studying the transition layer below the olivine layer below the asthenosphere, he only makes an associate professor's salary. He has been a full professor for 20 years, but not a great grant getter, and the department, the college, and the university reward most those who bring in the most money.

Fred is writing grant proposals. The department and the college have no funds for new research. The National Endowment for the Arts (NEA) turned down his request for money for a multimedia and specimen session at the national meeting of the Geological Society of America. Fred's collection of deep-sea material from Miocene strata in Argentina — the specimens, the serial sections, and the photographs — seemed to them possibly offensive to nonspecialists who might wander into the exhibit hall at the convention.

"This is my art . . . and science," said Fred. He also told the NEA it was unlikely a nonprofessional would be roaming around the meeting hall in Dallas since there would be signs, roped hallways, German shepherds, and armed guards on the doors to keep out those who had not paid the $200 registration fee. The number of people off the street willing to pay 200 bucks to view a coprolite show will always be few, said Fred. But the NEA would not reconsider.

Fred has now gone to the National Science Foundation. His brother, a lawyer, and his wife, whose passion is linguistics, helped him struggle through a program announcement — an especially capricious excursion into rhetorical secrecy — and there is money in Washington for coprolite research.

Apparently there has been funding for coprolites for years. An acting division director whose hobby runs to Pliocene mammal droppings has a private collection that he brought to Washington in a U-Haul trailer, spilling half a dozen boxes in a mishap in Indiana and getting in a scuffle with a farmer. "This weirdo was scooping up sheep droppings from my pasture by the Interstate," the farmer told the sheriff. The government paid for the trailer. They settled with the farmer.

The acting division director likes to get up in the morning before his wife and put spectacular specimens in the kitchen by her chair or beside the corn flakes. When he left a spotty brown trail from the bathroom across the new cream carpet to her dressing table, he was banished to the spare bedroom. The acting division director has supported the research of a favored few who search the Nebraska badlands for such significant mammal remains.

Two small groups of coprolite workers chosen by the National Science Foundation (NSF) read the grant proposals. The team is selected with care. They are the most knowledgeable specialists in the world, even the universe. They send faxes back and forth, call each other, and hold meetings at conventions. They hole up in obscure Virginia hotels. The NSF pays their *per diem* plus a stipend, fetes them at banquets and cocktail parties and pays airfares. They send all of their articles to each other and to 300 innocent people they believe have some influence in geology or an interest in coprolites, or who might someday reach such a state. The deep-sea specialists never mingle with the continental specialists.

So far, Fred has been unsuccessful with both groups. He has put aside his studies of continental coprolites. He has turned completely to the deep-sea varieties, paying his own field expenses to get a start. He has been interested less in fame than in being in the field, less in money than in scholarship, and the NSF has been willing to let him muck along without the corrupting influence of financial support.

In an evaluation of Fred, after looking at his most recent proposal, a reviewer concluded that he is not overly productive in research, crediting him with 20 papers during 30 years of publication. The reviewer piously noted that abstracts were not included in his assessment, because he regards them as meaningless. The reviewer generously allowed that Fred had attained a world reputation, but was forced to impugn the soundness of the world's judgment. The reviewer urged Fred to publish more and to get more grants.

In looking at his work, the reviewer somehow missed 150 of Fred's papers and all of his books, essays, abstracts, and book reviews. "This is a person I would not want counting lambs," I said to Fred, who also has rustic origins.

"An understandable mistake," said Fred. "They read scores of proposals."

"Yeah," I said.

Although some of Fred's critics note that he has published articles jointly with various colleagues, they say most of the work was done by others, presumably Fred's co-authors. What leads them to this conclusion is a mystery.

Fred remains a lot more optimistic about changing the direction of his studies than I do. He does not belong to the deep-sea coprolite club.

"These people are a nation," I said. "They protect their boundaries. They want to become more important in the coprolite world, but only if they can do it without granting citizenship to alien scientists. They cut competitors down, which they believe elevates them. They will do anything to maintain their hegemony."

"Prune that metaphor!" said Fred.

On Fred's Argentina proposal, a critic suggested sending a southern Argentinian north to do the study, although the Argentinian has no experience with the rocks, has not seen the fossils, and was unaware of the rich and varied remains in the formation until Fred wrote one of his meaningless abstracts. The Argentinian is working on other problems, has not proposed a study such as Fred's, and is known to be a researcher who needs to be led around by the hand in the field. According to his critics, any club member is preferable to Fred.

Before they recommend support for Fred, his critics expect him to attain a professional level they have not reached. They say they want him to be able to recognize coprolite species in the field. Fred could get around this by describing and giving new names to the species he doesn't recognize, an approach some of his critics take, but Fred believes in bringing specimens back to the lab and comparing them with other material. Apparently this procedure, followed by paleontologists for more than two centuries, does not appeal to the perfectionist wing of the deep-sea coprolite club.

The deep-sea coprolite specialists' efforts to squash Fred have lately turned to criticism of the references he includes in his proposals (7663 in the last one). The number of references to the specialists' work is never enough. Perhaps Fred is suffering a loss in episodic memory during his middle years. He forgets two or three critical citations, always to articles by one or another researcher — always a different one, depending on which anonymous reviewer reads Fred's proposal. "You've had some insecure jerks reading your proposals, Fred," I said, but Fred demurred.

Some claim that Fred incorrectly identifies what is critical, significant, seminal, important, and yes, of genius level, say his critics. The number of geniuses engaged in deep-sea coprolite studies is large, each only slightly less gifted than the one writing an appraisal of Fred's latest proposal.

There is certainly a great temptation to think Fred's troubles are compounded by his life-long interest in rocks. His critics believe this is old-fashioned of Fred — nonmathematical. They say he has not developed a feeling for the small convoluted picture, nor for the miniatures now much admired in coprolite research.

Fred discovered all the continental coprolites he has written about. He has discovered some new deep-marine coprolites. He has collaborated with another nonfunded nonspecialist on an article about their morphology, the rocks that contain them, and their ancient setting and history. Fred has linked the new coprolites to their probable maker. He is able to show, with photographs of specimens that span about 5 million years, that the makers were less constipated in the Pliocene than they are now.

The article has been accepted by one of the prestigious paleontologic journals. Fred gave an enthusiastically received talk at an international meeting of sedimentologists at Bologna. When Fred applied for his last grant, he sent a copy of the article to the NSF program director to demonstrate his talent for that sort of research.

How innocent. How naive. The program director chooses the reviewers and appoints the review panel. The program director did not call Fred, write him, or return the manuscript. Fred called the director once — his call was not returned. Fred would have had a better chance of being read if he had put his article in a bottle, tied it to a hunk of driftwood, and plunked it in the Great Salt Lake.

Through all of his tribulations Fred has never given up hope. A few weeks ago a reviewer gave Fred's newest proposal an excellent rating, saying it was the most comprehensive study proposed in many blue moons, yet credible and lucid. The reviewer suggested the research objectives were on paleontology's cutting edge, and the objectives might even be unrealistically comprehensive in scope. Fred was delirious. He waved the NSF form around all one morning, then put it on his desk with the rating turned toward visitors.

Meanwhile, like all of the others, the new proposal was rejected. The reviewers admired it, but said Fred was not a specialist in deep-sea coprolites. "Most of what is wrong with coprolite studies stems from the work of nonspecialists," said a reviewer who rated the proposal only good.

The critics for the NSF have deflated Fred on deep-sea coprolite studies. I am surprised he hung in so long. He had no chance, but he stuck until several critics claimed he was overqualified for such studies.

What could Fred do? He was always off the track. Those years of going to the field, making discoveries, writing meaningless abstracts, giving talks on deep-sea coprolite horizons, going to meetings, and putting together an occasional piece for the general public on his research had at last made Fred overqualified. Or so some thought. Worse there were no coprolite funds for over-qualified nonspecialists.

Fred has rebounded. A few days ago he sprang into my office, pulled out some snapshots of Cretaceous sandstone and asked me to look at the upper surfaces which, he pointed out, were embossed with peculiar sinuous ridges several centimeters wide and several decimeters long. The features superficially resembled some deep-sea coprolites. The enteroform structures were made up of sandstone much like that found below them — most were extensions of it. "The sand was injected up into cracks," said Fred. "Perhaps there was an earthquake and up it came."

"Pseudocoprolites," he said at last.

Shortly afterwards, Fred began writing grant proposals. He sings in the hallways. He is planning a trip to Morocco. The old fire is back in his belly.

Bestowing Glory in Lafayette

The best part of the morning paper, the sports section, isn't the same in a strange town. The columnists aren't the same, teams differ, and the scores aren't arranged in the same way. Breakfast in unfamiliar surroundings — I'll call the town Lafayette — seems not to be quite right either. The cantaloupe is green, the poached eggs mobile and fluffy, the tea scalding and, later, cold and too strong. The August heat and humidity begin early in the morning in Lafayette.

I would order fresh tea, but I must brush my teeth, find the meeting room, and prepare to record. At the last meeting of the awards committee, I was appointed recording secretary, an action, I believe, meant to give an impractical professor something to do while managerial types — the rest of the committee — make hard decisions. I find the meeting room and pull out yellow pads, assorted pencils, three fountain pens, and a tape recorder. There's a lot of nodding here and there and a satisfied glance from the chair, but, hell, anyone who knows me knows that I won't get the recorder working and that my writing is worse than that of a doctor who's written a penicillin prescription for a strep throat that may turn out to be the residue of a strawberry popsicle.

For several years the committee has tried to establish a system for ranking (1 through 12) the honors they bestow. They didn't anticipate the difficulties. The mimeographed lists help, but not all of the committee can recall an award when it is discussed. And then the lists change. Last year's Distinguished Service Award is this year's Outstanding Service Award.

Extinct awards are also a problem. Through the purchase of lavish medals (the association's emblem in 14K gold engrailed at the edge) and cash awards, the principal for the J.R. Hatcher Memorial Medal expired soon after J.R. toppled over at his club. Mr. Hatcher's widow — though driving a Cadillac and depositing Exxon dividends at a checkbook-rattling rate — views the funds as one of J.R.'s least admirable vanities. She is unwilling to continue the extravagance — an award, she asks incredulously, for speaking at scientific meetings? There's already too much of that going on in the universities. Still, the chair, a determined sexagenarian from Sioux City, insists on hammering out the ranking.

Unanimity prevails regarding placement of the Monroe I. Shackleford Medal. The medal is the oldest award of the association and its most prestigious. Tradition coincides with the committee's desire to keep it number one. The last recipient, however, attained his eminent position in industry with an unblemished publication record. Nary an article, not even an abstract or book review, mars his vita. That is unexceptional. His ascension to a corner office with a tiled bathroom and a jiggly secretary might well have stalled with the encumbrance of published research. The significant research criterion of the Shackleford has often given slight pause — but only slight — in considering candidates for the prize. Someone always recalls that Mr. Shackleford's research was also trifling, and it was his company reports and drilling successes that caught the eye of management at Goliath Hydrocarbons — oil, gas, tar sand, gilsonite, wurtzilite.

To everyone's surprise, the committee ranks the awards with little argument. The Shackleford at 1, Honorary Membership at 2, J.R. Hatcher at 3 — provided funds are obtained — National Service, Outstanding Service, Director's Award, and relentlessly on to 12, the Peerless Plaque. The order will be publicized in the journal — most prestigious printed first — and at the annual banquet — most prestigious presented last. "A nice piece of work," says the chairman.

The next order of business — a proposal to limit honorary memberships to 50 — comes as the coffee and doughnuts arrive (late). The sponsor, Mr. Michael Bloodroot, says too many holders devalue the award. Enough sexagenarians and septuagenarians, still hopeful of recognition, are sitting around the room for that notion to be acclaimed. Two push their chocolate doughnuts aside. Mr. Bloodroot reminds the committee that Montaigne, writing of the Order of Saint Michael, warned that to annihilate an honorary award you only have to lavish it on multitudes. "There are a lot more deserving people in the association than there used to be," pleads a septuagenarian. "The honor shouldn't be degraded *just* for that reason," says Mr. Bloodroot, citing Montaigne again and, for the clincher, throwing in the 1st-century Roman epigrammatist, Martial. The executive director says 46 honorary members are alive or nearly so. A move to table the motion fails. The vote is close. The roles will be held to 50 honorary members. Two years ago, at the time when there were 42 honorary members, Mr. Bloodroot was one of those elected.

"Shall we raise the Director's Award stipend to a hundred dollars, lift the age restriction, and just give it to the author of the best article in any association publication?" asks committeeman Jim Billy Smith. Before my thoughts wander to ancient lakes and streams and yesterday's tennis, I try to focus on the endless questions: "Is the action retroactive? Can ... Will ... Jim ... J.B. ... ?" I hear the chair say "passed." Lunchtime.

The buffet: prime rib, ham, pork, mashed and scalloped potatoes, corn and peas and carrots, three gravies, fifteen salads, four salad dressings, five nonalcoholic drinks, and cheese cake with strawberry or blueberry topping. Flowers — roses or carnations indistinctively fashioned — adorn the crystal. A simply designed oil derrick — red enamel superstructure outlined in black on cream — crowns the plates. Sturdy silver, tastefully engraved with the club's initials, borders each setting. Committeeman Rob Drake says "We've got to get Washington out of the oil business. They don't belong in the private sector. Never have. Never will."

The chair, after brandy and coffee, maintains that the banquet lasts too long. Older members drowse; their wives fidget. Last year an honorary member fell asleep following the Peerless Plaque and missed everything between that and the Shackleford. (His nose barely missed the apple cobbler.)

Jim Billy wants to continue having all association awards at the banquet, but declares that eliminating awards for one division is viable. ("Yes, quite viable," he repeats briskly.) The committee decides to continue the full program.

The air is intoxicating — the smell of medals, plaques, fame, and stale cigar smoke. Heat also permeates the room — the afternoon sun and humidity are barely relieved by the air conditioning. After wallowing in so much glory, the thought that I might someday be honored gallops into a tired mind. I would like an award. A small award, say number 12, which is given for insignificant but diligent and long service, seems about right. I've been the consummate committeeman, serving the longest continuous term in the Assembly of Delegates, unmarred by introduction of any legislation. Years spent on the stratigraphic correlations committee, undistinguished by proposal of correlations. Hours in dark rooms at conventions, judging papers and slides. A term on publications, counting dry holes drilled in Utah. And, yes, other assignments as well. There should be an award in all that.

The committee grinds inexorably on, but now from memory and momentum. Everyone wants to finish quickly; there's a dinner this evening. Finer prime rib than that at lunch is promised. Hands move to shade eyes, a sure sign that eyelids are closing furtively. Committeeman Buddy Houston, not content with naps, pulls colored glasses out and, with the garish dash of a Glenn McCarthy on the rise, puts them on. The strawberry cheesecake and brandy react acutely in the secretary. Momentarily, I wish that I had worn an eye patch or two eye patches and a hood. The dizziness passes. Committeeman G.I. Beatty (Gib), through decades of committee work, has learned to nap with his eyes open. His arm jumps involuntarily, however, and his crony, Gabe Aston Bayne (Gab) spills his coffee. The chairman cites the Journalism Prize: "No. Not this year," he says.

That prize honors the author of a book or several pieces that contribute to public understanding of the earth or the oil business. Giant oil fields apparently are commoner than fine writing. A trade journal reporter is the only one proposed. "It's his job, isn't it?" the chairman reminds us. No argument.

The pace quickens. The committee gets its second wind. The smell of dinner works its way into the air conditioning. The

Director's Award goes to an exploration manager who found gas under a refinery — "... imaginative yes, and a blockbuster exploration program." The Peerless goes to the vice-chair of the meeting in Anaheim — Disneyland tours and a trip to see the sinking city, Long Beach. The National Service Award is given to the man who organized and hosts the annual Galveston fish fry for Sea Scouts. An honorary membership falls to a former president, one of those sitting along the wall. The committee revives and grows festive. "A fine day's work," the chairman says, and invites the committee to his suite for a little libation. I decline with hearty thanks and dump everything but the recorder — undisturbed throughout the day — in a five-dollar plastic briefcase.

Praying and Fasting Over Great Salt Lake

Utah Attorney General David L. Wilkinson has declined to answer — at least for now — a letter from a Wisconsin-based group protesting some Utah County Commissioners' plea for a day of prayer and fasting for lake flooding relief.

The "Freedom from Religion Foundation" of Madison, Wisconsin, wrote Mr. Wilkinson 10 days ago calling the commissioners' plea an "unacceptable abuse of separation of church and state."

— Salt Lake Tribune, May 14, 1986

Last spring when the Six County Commissioners' Flood Task Force organized a day of prayer and fasting for lake-flooding relief, our little group didn't show much interest at first. It took them two days to remember that we have many specialists on Great Salk Lake in our department — as it turns out, almost all of our geologists and all but one of the geophysicists.

Once they realized this, it wasn't long before the meetings began. Our faculty began to talk at the student union, on the elevators, in the hallways, and in the offices, sometimes behind closed doors. We met a little above halfway between the Stansbury Level of ancient Lake Bonneville (at 4470 feet) and the Provo Level (4860 feet). It soon became evident that a few of our people had a lot to say, at least among themselves, and that some had ties — cousins and uncles — to several of the six county commissioners or their wives or cousins or aunts or uncles.

At the beginning of the first flurry of meetings, I was uncertain what my course would be on prayer and fasting — either, neither, or both. I have not been successful at prayer. I believe I have always been too specific — a date with Patty, a C in Physics, a .22-caliber rifle when I was 12. And the spread of my waistline attests to how few days of fasting there have been. I also began to think my participation might be unwelcome, an impurity in the pureness of the other participants. The commissioners were never clear on how ecumenical their plan had become or was intended to be.

Professor Albert Bloodright, although like me an unlikely candidate for the fast, claimed to have all the lake-data at his fingertips, which were stained black by clippings from the *Salt Lake Tribune* and brown from $25-per-pound Belgian chocolate he now buys by the box. The lake level, he said, is at 4211.01 feet above sea level, barely below the recognized historic high of 4211.6 feet set in June of 1873. The lake's 11.6-foot rise since the 10th of October, 1982, represents a historic rate of increase. Last century's climb to that level took 12 years to accomplish. And so on. "The old lake-level record will be thoroughly smashed," he said with satisfaction.

Bloodright's crony, John Preston Worth, questioned the 1873 record. "It might be off by as much as a foot in either direction," said Worth.

Records of the fluctuating level of Great Salt Lake were not begun until 1875. G.K. Gilbert, one of that era's most famous geologists, made the 1873 estimate on "circumstantial and traditionary" information.

Professor Johnson, recently tenured, much to Worth's chagrin, said it didn't make a lot of difference whether Gilbert's estimate was a record. When the lake level rose week after week, month after month, it worked disturbing miracles. Familiar objects disappeared and it was easy to forget they were ever there. Roads slipped under water, surviving only in a sign instructing the unwary: DO NOT ENTER. Storm waves coursed through a never-opened Saltair Pavilion, carrying sediment into dark dance hall corners intended for lovers.

I remembered burying my dog Ranger, a victim of his zest for moving car tires, in an oolite bank on the then-south shore. The road to Ranger's grave is gone, the fence posts beside it are under water, and the lake has covered Ranger with 15 feet of salt water. I can visit his resting place only by boat. Sometimes I row out and drop a Sego lily about where I think he lies, hoping it will float above him.

Velma Holmes, the second woman hired on our faculty (now 8.7% female), told me that a lake-shore boulder she had often sat on to watch the evening light has disappeared, covered by the rising lake during the April runoffs. She and her boy friend were in the habit of going there on Sunday afternoons to enjoy the lake and read Philip Larkin and James Wright. All of the large rocks along the south shore north of the Oquirrh Mountains have been submerged, she said.

The rain and late spring snow came every day for several weeks in April. The dikes were reconstructed, but a fierce storm broke through them and flooded Interstate 80 and the salt ponds. Union Pacific dumped trainloads of sand and gravel along low-lying sections of the tracks, which act as a breakwater for Interstate 80. Winds that reached 70 mph drove waves against the new dikes, eroding both sides and eating into the structure until the dike broke again. Professor Holmes said the breach did not open all at once.

The chair became interested in the rising lake and the day of flood prayers. "Feeling is running high that something has got to be done," he said. "And we're losing the chance for grants. Let's focus. We can still get in there."

The chair earned his degrees at our place and has been much honored by it. His first internal award led to passing of what Johnson calls the Cosworth Rule: No awards to anyone serving on the selection committee at the time the selection is made. I'm sure Charlie was as surprised as everyone else at the committee's decision to honor him with the research prize. After all, he was chair of the committee and therefore may have known little of what the subcommittees were doing. "But he didn't refuse it," said Johnson one day.

Charlie has the strongest connections of anyone in the department with the newspapers and the television stations. Every few weeks The Chronicle (Chrony for short), the school paper, carries an article on one of our geologists. Our biggest stories come from the seismologists and the small group interested in the dinosaur's demise — comet collision or shallowing of the oceans. Charlie can always persuade public relations to push any news of his or our adventures.

At a later meeting, he asked if we should form an action group. There has not been a lot in the papers on the lake attributed to any of our people. There was a suggestion to drill a well by the Hogup Mountains on the west side to see if the gravel aquifers are dry. Some of the water might be drained off into a drill hole: "Pull the plug on the lake." But interest in that idea faded quickly. "This fellow Ted Arnow with the U.S. Geological Survey is the one getting all the press," said Charlie.

We talked for 20 minutes about the task force and decided against it. On task forces there are apt to be but few workers and many advisors. We were unanimous in the opinion that we did not need a new committee.

"Would anyone like to meet a commissioner?" said Charlie. We were unaware that Charlie's wife and the wife of one of the commissioners had been schoolmates at Provo or Nephi and are second cousins. I spoke against that. There was a county commissioner in my family, and I knew how long county commissioners can talk and how short are the pauses in what they say. When they get going on whether they need to buy a new road grader this year or can wait until next year, time passes geologically.

As it was, the new numbers Professor Bloodright had got hold of threatened to carry us on into the next dry cycle of the lake. "The rise from 4204 to 4205 feet cost $55 million. Total damage is more than $200 million," he said. "A rise to 4218 feet would endanger the Salt Lake International Airport and if the freeway has to be moved south, there may be $400 million in rebuilding costs. The lake's area has increased from 1640 square miles to 2450 square miles, which is larger than Delaware."

At our last meeting of all the faculty, Bloodright coughed and wheezed — his allergies — through the details of the proposal to pump water into the west desert. Governor Norman Bangerter, though agreeing that the plan has many and varied faults, suggested that no other reasonable alternative exists. Some plans cost too much, others do not remove enough water. Bangerter has said little about prayer and fasting. "West desert pumping is the best choice," said Bloodright, a supporter of Bangerter in the last election.

The pumping may reduce the lake's elevation 14–16 inches the first year and 7–8 inches each subsequent year.

Some critics have suggested that if the lake rises another 4 feet, water will flow naturally into the west desert. Pumps would be useless. Bloodright and the Governor disagree with the suggestion.

The state might also be left with its pumps high and dry. Inflow into the lake next year will have to be 160% of normal to keep the lake at its present level. "Anything less, and the lake will go down on its own," said Johnson. Almost everyone but Bloodright recognizes that expanding the lake's surface area will produce more winter fog. This will enhance the lake effect and increase precipitation along the Wasatch Front and also increase inflow into the lake. So it may go.

The pumping project will cost $55 million or $70 million. Opinion differs. Bloodright put half our group to sleep with the details. The three pumps will stand 50-feet tall and have an output of 420,000 gallons a minute. They will be made from corrosion-resistant aluminum–bronze alloy. Each will have an 11-foot-diameter discharge tube. The project will create a 315,000-acre evaporation pond west of the lake. Annual evaporation will be an estimated 820,000 acre-feet. "I wish he would read Thoreau instead of the *Salt Lake Tribune*," Johnson whispered.

Johnson could finally take no more of Bloodright's numbers or ideas. "These people are dealing with a three-dimensional problem with two-dimensional minds," he said. He reminded Bloodright of the cost of breaching the causeway between Promontory Point and Lakeside and how little the lake level fell afterwards. "West desert pumping looks like another fiasco," he added. "They'll have plenty of trouble even getting the pumps over there."

Bloodright spluttered. Johnson said the plan had the look of $55 million wasted, probably closer to $70 million. "And what will happen if west desert water runs right back into the lake?" he asked.

The meeting broke up. We were on our own. Charlie couldn't see much money or glory in lake control. It could backfire. There would be no task force or action group or further meetings among our faculty. The chief topic of conversation was no longer the weather, the flooding lake, and the west desert pumping plan. Quarreling over space, teaching assignments, committees, and salaries resumed.

At sunrise the morning of the official prayer and fast, I parked at the lake shore west of Black Rock Beach in a view area on the north end of the Oquirrh Mountains. The sky was slate-gray and the wind began to rise. Waves on the south shore broke against the Union Pacific dikes. Salty spray rose above the tracks and fell in rhythmic showers on the rails and over the gravel beaches. Spring seemed to be coming in again strongly as it had for many weeks and for several years.

A California gull (*Larus californicus*) walked among the debris of three overflowing trash cans, picking something up here, passing by something less edible there. In late February or early March they lay their eggs, and then raise their young during the summer. They return to California in late July or early August and spend the winter there.

The gull moved to the top of the terrace above the cans, its head darting this way and that. From there its view of the lake and me was greater, but it seemed unconcerned about anything. I imagined the gull considered me harmless. It was prompted to begin a series of squealing calls: *kiarr . . . kiarr . . . kiarr.* I couldn't tell if these were for sunrise, for a distant mate, or in delight at a fine breakfast. Its voice filled the air with song until a car turned into the parking area and stopped below the terrace.

I fasted through a day and prayed the state wouldn't meddle too much with Great Salk Lake. The lake rises and falls and pays no attention to human structures or opinion. When the first permanent settlers came to Utah in 1847 the lake level was at about 4200 feet, perhaps a sort of average for historic time as Professor William L. Stokes has suggested. The lake doesn't linger at any level for long.

The storm broke soon after sunrise and it rained and hailed off and on all day. Snow fell lightly on olive-gray limestone of Oquirrh peaks. A good-looking, middle-aged woman standing next to me said there had been another day of praying and fasting in 1981 during a dry spell in the valley. Those prayers had been answered almost at once, she said.

In the Field

Mysterious Maverick Range

The Uinta Mountains stretch in a long line; high peaks thrust into the sky, and snow fields glittering like lakes of molten silver; and pine forests in somber green; and rosy clouds playing around the borders of huge, black masses; and heights and clouds, and mountains and snow fields, and forests and rocklands, are blended into one grand view.

— John Wesley Powell, 1869

Since the spring of 1951 the Uintas have never been far from my mind. That was the year of my first crossing. Soon afterwards my job would be to study the basin south of the mountains. But one must understand the mountains in order to understand the basin. They go together yoked in a marriage that will last until the mountains wear away.

I've crossed the Uintas or traversed their length perhaps 40 times, each trip different. The mountains never look the same, and the little I learn each time accumulates slowly. The mountains hold their secrets tightly, locked in crystals and grains. Unlike most other ranges, the Uintas run east–west. To further confound the world, they've moved north 12 miles into the northern basin — maverick mountains loose in the crust.

A few days ago we — Bruhn, Picard, Beck — finished a manuscript on the history of the Uintas and on the ancient geography of the southern basin. Printed on an Apple computer in a corner of Bruhn's basement, the words are black and blocky, the tables sharp and clear, and the figures glossy and substantial with confident, thin, dark lines that trace the evolution of mountains and basins. Are we close at all to the actual history? The manuscript is on the editor's desk. Its absence leaves me both apprehensive and relieved. It's too late to do anything more with the words. I want it back and I don't want it back.

On this bright February morning I decide to look again at the
Uintas. My friend, McBride, is in town and wants to see the
range. His nerves are unsteady from administration duties; mine
are just unsteady. There is a chance that on the summit or on
one of the flanks, nerves will quiet. Geologists have always gone
to mountains for sanity and safety.

McBride has been up since 6 o'clock, sitting in the hotel lobby
reading a whimsical piece about the bestowal of professional
glory in Lafayette. He rises early, has high energy, and is always
ready for a field trip. He is short and trim, has a prominent Adam's
apple, and has brown curly hair that further softens an already
boyish face. The first time I saw him, he was standing in a hotel
lobby in St. Louis, talking to a student about a thesis. A crowd
of listeners gathered around him. Some of them were joking
about his bolo tie whose center piece was a scorpion encased in
plastic. About 20 years ago he owned a single club tie. Bolo ties
were his trademark then. Now he wears club ties and dark blue
suits. Geology is his business and, except for an occasional movie,
geology is his recreation. His specialties are sandstone, West Texas
formations, and chert. These have been enough to fashion an
illustrious research and teaching career. His bibbed overalls, red
flannel shirt, and pink sweater take him a long way this morning
from club ties and blue pinstripes.

We set out late. My fault; I overslept. To get to the north side
of the Uintas we cut through the Wasatch Range east of Salt Lake
City, cross young, 30-million-year-old volcanic rocks, and turn
northeast between the orange conglomerate walls of Echo Canyon — Interstate 80 to Evanston in southwestern Wyoming. Drift-
ing wood smoke, gray dirty clouds, and oil derricks hang above
Evanston. The light snow during the night has melted, and the
hilly streets above the Bear River are wet and slippery. Evanston
resembles other large towns along the Union Pacific railroad —
Green River, Rock Springs, Rawlins, Laramie. Evanston is the
county seat of Uinta County, the 14th most populous county of
the 23 in Wyoming, a state of few counties and few people.
People in Wyoming like it that way.

McBride wants to grab a sandwich. I hesitate. I scan my memory and find nothing on edible sandwiches in Evanston. From

the corner of my eye I see JB's. We turn sharply right, pull alongside a gas station (soon to become a drive-in bank), then make
another right turn onto a muddy dirt road into JB's parking.

The waitress deposits the tuna sandwiches. The bread is awry,
revealing old lettuce that is brown-speckled and wilted. I gently
raise the lettuce and find hard, dark tuna chunks astride a thick
paste. I reassemble bread, lettuce, and tuna. Jack Nicholson or-
dering a sandwich in *Five Easy Pieces* comes to mind, but I can't
carry that off. I eat the sandwich.

We set off again — east on I-80 to the badlands. The nearly
flat-lying sandstone and mudstone accumulated in ancient streams
and on floodplains at a time when water was abundant. Now
there is little water and little soil or vegetation.

The drivers along I-80 seem frenzied. They fight the barren
desolation of rocks that must appear relentlessly monotonous to
them, the gentle curves of the basin floor, and the great distance
ahead to the next town—Evanston to Green River, 88 miles.
Green and gray, maroon and mauve, buff and brown, the sandstone is carved into castles, ancient totems, and Roman pillars.
Rare rains have sculpted dragons, dinosaurs, and gargoyles. I see
friendly faces in the sandstone, but also mugs that could stop
your heart at night. Cut by water pouring down draws, and polished by wind, rock images are commoner than cars once we
turn southeast off I-80 at Fort Bridger and cross Black's Fork.

We travel east. McBride sleeps. The sun is high and bright,
uncovered by white clouds coming up from the southwest. The
air is warm and sweet. Scents of wet clay and scrubbed sandstone
rise from small patches of melting snow. A spring day in February,
a lucky day for this time of year in the badlands. All of the figures
etched under the snow during the winter are uncovered in the
basin.

We cross sediment of cold streams flowing north away from
the Uintas — Black's Fork, Smith's Fork, and Willow Creek.
Black's Fork was named for Daniel Black, a trapper who was one
of General Ashley's men; Smith's Fork was named for Jedediah
S. Smith, another mountain man who was in the vanguard of
trappers who worked the beaver streams on the north flank. We
turn south and head for Mountainview. From here the long,

nearly unbroken crestline of the Uintas looks fretted. Mountain glaciers have carved the crest, cutting amphitheatre-hollows and leaving knifelike divides between the valleys — the rugged topography of horns, cols, arêtes, and cirques. Only 10,000 years ago or so, all the valleys were filled with snow. The celebrated scenic mountain terrains — the Alps, Sierra Nevadas, Cascades, Rockies — were chiseled by glaciers. The jagged, serrated, linear crest of the Uintas softened by the winter snowfield and lightly covered by white puffball clouds is as fine as the crestline of any range. Driving along the basin floor and looking at the high peaks, one wouldn't expect to find a pass over the mountains, but there is a pass and, in February, one only.

I nearly run off the road looking at the rocks — a crazy geologist meandering down the road.

McBride wakes up. "A sinking spell," he says. "Where are we?"

I hand him a map, pointing to a spot between Mountainview and Lonetree.

We're well across the alluvium riding on gently inclined sandstone and mudstone. The stream beds are at least 35 million years older than the alluvium. Sedimentary rocks. McBride has studied similar stream strata in Texas. I ask him about the uranium in them. We compare notes. "Most geologists spend all their professional life looking at sedimentary rocks," says McBride.

The jobs are there because oil is there, along with coal, uranium, oil shale, tar sands, and even a lot of metallic ore deposits. The same is true for South African conglomerate which contains most of the world's gold in braided stream deposits formed in basins similar to this one, but twice as large.

Unusual among mountains, the Uintas are one large block of sedimentary rocks. The core is sedimentary, the mountain is sedimentary. On a geological map we find a few small dikes, but no other igneous rocks. Thousands of feet — once, perhaps 50,000 feet — of sedimentary rock crops out in the range and on its flanks.

The badlands remind me of moonscapes in old movies. The olive-colored rock is arranged in buttes and mesas and in slopes of dry draws that were cut by flooding streams. We cross a shallow brown stream. There is a little water moving drunkenly. McBride discovers a thin cream-colored layer cropping out in the east slope. It looks out of place between the olive beds. He wants to stop.

I slow down and turn into the draw. Small snow patches survive in the shade. Clear ice patches cover brown water in shadows along a meander bend. The east slope is bare and bright in the sun. A cottontail runs down the draw. We cross his trail lightly marked in snow.

The slope is steep, slippery, and sticky. Melt-water has soaked the ground and filled openings between grains. The expandable clay mineral smectite — formerly and, joyously, known as montmorillonite — has expanded, contributing its share to the slope's instability. Mud sticks and cakes and rolls up our boots. I turn to angle up the rise and, without warning, discover my feet in the air. I sit awhile. On a smectitic slope, the curved surface of a seat is remarkably stable.

McBride breaks off a chunk of the cream layer and looks at the grains with a 10-power lens. He licks the fresh surface and proclaims it limestone. He finds quartz and feldspar grains laden with biotite and hornblende — common minerals in volcanic rock. He kneels to press his nose against the cream bed.

As I struggle up the slope, McBride shouts, "Algae in the limestone. What do you make of that?"

Still below him, I don't know what to make of it. I am on my knees in the mud inspecting feldspar grains I can't see without ten magnifications and the chair of a prestigious geology department is dropping acid on a rock to watch it fizz. Humor him, I say to myself, and answer "Lake deposit."

Regaining my footing and my dignity, I answer more properly, "Volcanic debris fell or was washed into the lake. Likely both."

"Yeah. A carbonate lake." He pauses, still squinting through the lens. "Seems so." He breaks loose another limestone chunk and starts down the incline with it.

We clean our boots with sticks. Scraping is the only sound in the draw. The air is clean and warm and nearly motionless. There are no birds, not even a sparrow. McBride wraps the rock in newspaper and puts it in the trunk. We're close to Manila.

The weather's good in Manila. It lies in the rain-shadow north of the Uintas. The mountains break up storms. If more than a handful of people could make a living in Manila, it might become a city. Its nearness to the Uintas and water, along with its sparse population, has always attracted a few people, attractions that may some day put an end to its scant population. We stop at the general store and buy warm root beer, oily peanuts, and middle-age fig newtons. The store's paunchy owner, Orrin, fishes at Flaming Gorge. He also collects fossils, arrowheads, and scrapers, all of which are for sale. But mostly, the collection gathers dust in a showcase. Root beer and nuts sit on wooden shelves. Fig newtons are a sideline, stocked for tourists during the summer and, sometimes, still available in February for geologists.

McBride names the types of chert among the arrowheads and scrapers. Orrin shows us mammal teeth he's collected from the olive rocks north of town. The locality is a secret. I suspect he plunders it when the fishing's slow at Flaming Gorge. "Manila is a fine place to live," Orrin assures us.

McBride is eager to go on. We are closer to the mountains than we have been all day. Directly southwest of us and immense on the skyline, the western peaks rise nearly 7000 feet above our heads.

Utah-44, south of Manila, runs straight and fast at the up-turned rock in the low foothills. The gray-flannel-colored mudstone deposited in Cretaceous seas is barely exposed, covered in long east–west valleys by thinly spread vegetation and thin soil too poor to produce more than a meager living for animals and humans. On the other hand, the ancient Cretaceous oceans on this continent supported life generously.

The road turns to switchbacks and rises steeply. The rock is mostly red hued — light and dusky red, brown, pink, and orange. McBride looks at the map and makes a short list of places: Red Creek, Red Mountain, Red Knob, Red Canyon, Red Castle Lake, Flaming Gorge, Vermillion Creek Basin, Brown Duck Basin and Mountain. The pass is bare, but the air feels cool. We smell snow in the light breeze. Spring is in Manila and winter reigns in the mountains. July will be here before the evergreens are free of the snow banked against their trunks. On the high peaks, cirques are snow-filled and white the year around.

Each switchback swings us crazily across the north flank —
like a ride in a kiddy car at the carnival. Across, down, around,
and up. Beds are on end, overturned, and cut by east–west-trend-
ing faults. I suddenly see a major north-flank fault. The mad
gyrations of the road and the chaotic, steep, overturned beds had
partly prepared us for its sudden appearance: a gigantic yellow-
orange spire of Mississippian limestone stretching skyward, but
tipped northward carrying us back into the mountain below
blood-red Precambrian sandstone. The most ancient rocks have
moved over younger rocks in the north-flank fault, a reverse fault.

"Wow!" McBride says. We stop and walk to the top of the
slope. The air is cool and a misty mountain wind moves down
the valley. We're below the snowline, but winter is near in the
shadow on this west-facing slope. McBride zips up his pink
sweater and pulls on a coat. Thin Austin blood. We photograph
the fault and the beds south of it. To McBride's amusement, I
take my pictures twice. No film the first time.

We traverse six switchbacks and rise into drifted snow and
blood-red sandstone. It is the core of the range, outlining on the
map the approximately 150-mile length and 40-mile width of
the Uinta anticline, the colossal Uintas upfold. The rocks are
appropriately called the Uinta Mountain Group. McBride wants
to sample the sandstone. I pull over, but long snow windrows
protect the sandstone from geologists who would whack a pick
against sandstone faces. I've sampled this place before. Not today,
however. Snow banks lap in long tongues onto the red beds and
fence the rocks from the road. A coyote has run up one such
tongue. We see rabbit tracks, the only other sign of life. This
sandstone is more than a billion years old. The record of inter-
vening time between the coyote and the billion-year-old stream
that deposited the sand grains has been eroded away or was never
left. No record of the lost time exists. Coyote and rabbit and
sandstone do their dance.

There's plenty of sun, but west and south slopes are dark and
chilly. We've crossed the narrow north flank and are well onto
the broad crestal plateau of the large fold. McBride stares at the
snow. Snowstorms are rare in Austin. When it snows, school is
canceled; and while the snow lasts, kids make snowballs and

adults bump bumpers. This Uintas snow is banded with blood-red sandstone stripes and groves of dark evergreens. Green, white, red. Like the Italian flag, McBride says.

I hand McBride cross sections. I think we're near the mountain's center, but am uncertain. The Uintas moved northward 12–13 miles on a major, north-flank, thrust fault that is inclined southward to a depth of 9–12 miles. The crust beneath the southern basin was transported eastward 12 miles relative to the northern basin. Jostled crustal blocks. Or so we think. Geologists sometimes call such drawings cartoons, especially those drawn by someone else. Our rambunctious drawings of this mountain and the basins are distillations of surface and subsurface studies. Imagination, too, McBride adds. I hand him the fig newtons. He wisely declines. I do not.

McBride wonders how our Uintas theory fits plate–tectonic theory. "There are other east–west ranges on the North American plate," I say, trying a transparent diversion.

Our feeling, and that of many other geologists, is that the thick Uinta Mountain Group is the fill in an east–west-running trough — an aulacogen — that extended into the continent during late Precambrian. The position and orientation of the Uintas is mainly controlled by the ancient trough and by associated crustal faults along its edges. The range forms an enormous anticline that has been pushed northward and shortened up a fault ramp that, in turn, extends downward into brittle–ductile rocks. Movement began about 65–70 million years ago in latest Cretaceous time and continued intermittently for 30–35 million years. What crustal block hit the trough, we do not know. Ideas on that subject are wild, a subject for another time.

McBride puts the cross sections down and looks at the geological map. He nods. "The figures hang together," he says. He picks up the cross sections again.

McBride asks about other drawings. Two triangular diagrams that represent sandstone composition catch his eye. I recall Harvey Blatt's explanation: "The popularity of the triangle is probably attributable partly to its value for graphic display and partly to motives that only a psychoanalyst could plumb."

McBride wants to walk in the snow. We stop near high thick evergreens that surround an elliptical clearing. From the forest's edge, the wind roaring in the treetops sounds like breakers along beaches of an ancient shore that carried uncountable grains from the Uinta Mountain Group down and along shallow marine slopes.

We take several steps into the clearing. Olive-gray clouds nearly touch the treetops. The only sounds are wind and rustling trees. McBride takes pictures of drifted snow, trees, and an island of rock. Droplets of water appear on his camera and run down McBride's glasses.

As we enter the south flank switchbacks, the silent sun appears low in the west. Shadows stretch out from trees and younger upturned rocks. We pass yellow-orange Mississippian limestone, and a thick sandstone.

McBride thinks he sees the Nugget Sandstone. He is right. He wants a sample.

He raises his hammer and wallops the Nugget. We're standing in the last bright sunlight. Shadows cover the western slope. The cool air feels magical, parting gently, it seems, as I walk along the road. McBride stops, listens, then walks on, and whacks the Nugget again. McBride's third whack flushes a deer from hiding.

We wrap the sandstone and put it in the trunk. We go on slowly past Entrada Sandstone and marine Curtis beds. The dinosaur-bearing Morrison Formation overlies the Curtis in fine pastel-colored outcrops. We cross Ashley Creek, then pass a defunct pizza parlor and move on to Vernal — self-proclaimed capital of the southern basin.

At the Dinosaur Motel, they say a mountain lion was seen on the Green River near the Red Wash oil field. As unlikely as it seems for one to be so far north, I believe this to be true.

Tracemakers

they greet with illustrious fervor
each least, lost gesture of life

— Thomas John Carlisle

Donald W. Boyd is a paleontologist who, between expeditions, teaches at the University of Wyoming. He has made important discoveries there and in other Rocky Mountain states, as well as in Mexico and Algeria. In Algeria, he worked a week while suffering from pneumonia. He stuck it out long enough to make a significant find before flying home to Laramie to fight the infection.

We have come to Lander in west-central Wyoming to search for fossils in Nugget Sandstone. Remains or evidence of ancient organisms are rarer in the Nugget than rutabagas in Washakie County. Several organic trails have been reported, but just where they were seen and what they are no one knows.

The poor results are closely related to deposition conditions. The Nugget was laid down mainly in desert settings in the Early Jurassic. It was hot. There was little rainfall. There were small lakes within the dunes, and an oxidizing environment, which led to the destruction of the few organisms that lived near the equator in the sand seas.

Boyd has come from Ten Sleep in northern Wyoming. I drove in late last night from Salt Lake City. We ran into each other this morning on Main Street. If you want to meet someone in Lander, it doesn't take long if you walk on Main Street.

I ask Boyd what he thinks about some possible trace fossils my older son and I found in the Nugget 2 years ago. I describe them. He says he'll have to see them and let's get some breakfast.

Boyd is meticulous and prepared. He carries only the equipment necessary for collecting — no more, no less. Even so, his car is filled with gear. As we bump along in his cramped '75 Subaru, I think of Charles Smithson, the paleontologist in *The French Lieutenant's Woman*, searching for echinoderms in crumbly gray-blue limestone ledges. "He wore stout nailed boots and canvas gaiters that rose to encase Norfolk breeches of heavy flannel. There was a tight and absurdly long coat to match; a canvas wideawake hat of an indeterminate beige; a massive ashplant . . . and a voluminous rucksack, from which you might have shaken out an already heavy array of hammers, wrappings, notebooks, pillboxes, adzes and heaven knows what else." Nearly seven decades separate Smithson and Boyd: Smithson the methodical Victorian amateur; Boyd, the methodical modern professional. Under the skin they are close, closer than Charles Darwin and James Watson.

We pull off State-28 in the Wind River Mountains and park. I lead the way. I worry that I won't find the right exposures. The markings we're looking for appear when looking at the rocks in a particular orientation and when the light is right. The light is critical. The remains, if remains they are, may stand in relief for only an hour shadowed at the crucial angle. We might not see them at all on cloudy days. On an average field day, physical conditions can be right, and our eyes not sharp enough to see the marks.

What are the chances, really, of cutting again the ancient trail? The sandstone represents several million years of Earth history, a trail made in an instant. Then the wind swept most surfaces clean — an event that took minutes to happen caught in an eternity of rock.

We wander back and forth across the orange sandstone, foregoing systematic traverses. Trace fossils deserve their name here. Tracks, trails, burrows, and borings are more common than invertebrate body fossils in most sedimentary rocks. Rather than actual body parts, or casts and molds of body parts, trace fossils express the behavior of organisms, though they may be obscure as in the Nugget.

Small recent dunes, the color of old gold, sit on large ancient dunes. The small dunes may cover the aged traces we thought we saw. I scuff the sand away above a dune. No trails, but my boot is full of sand.

The sun is too high, I think — too much light. It is hot and dry. We may even be in the wrong place. Boyd asks again what they look like and what the proper perspective is. I am embarrassed. Trails that have lasted 190 million years surely survived a couple of Wind River winters.

We come at last to outcrops that feel more promising than those we've seen. I crawl on all fours, head moving left and right, neck protesting the quick movements. Loose sand sticks to my hand. My right knee aches. Crawling up the gentle slope of an ancient dune I raise up to see where I'm going, and catch sight of a trail on a bedding plane. The continuous groove disappears under the overlying bed. Before calling Boyd, now searching downslope, I study the mark and make sure it is what my son and I saw.

Boyd comes quickly when I call. He squats, puts his hands on the sand, and crawls slowly along the bedding plane. There are other traces. He pulls out his hand lens and squints at one. Usually the most talkative of men, he is silent. "Say something!" I shout to myself. I remember that he wrote a paper called "False or misleading traces." Various native features of sedimentary rocks, he noted, may be mistaken for evidence of behavioral activity by organisms.

Boyd takes a quarter out of his pocket and puts it face up on the bedding plane. "You've got something here," he says, focusing his camera and moving near the rock surface for a close-up.

With trace fossils one is almost always uncertain of their origin. Most were once thought to be fossil algae — fucoids. Sometimes drawings of the fossils were modified to more closely resemble algae. From his studies, Nathorst wrote in 1881 about the animal nature of fucoids. The article was never translated from the Swedish. Only in the last three decades has ichnology become a major discipline.

Boyd finds trails at many horizons. There are vertical burrows — excavations — but they are rare. We study, photograph, and measure trails and burrows. The mean length of long trails is 7.3 inches; width, 0.2 inches; diameter of burrows, 0–.4 inches. The average animal was headed N 83°E or at 180° to that direction — S 83°W. We can't choose between the two directions. The average wind direction was N 58°E. A burrower with good sense would head downwind, I say. To be cantankerous, this one may have hunkered down and moved upwind just below the surface, eating sand all the way. We think downwind is right, though. But maybe. . . .

The longest trails are oriented at N 11°W (S 11°E) — almost 70° to the ancient wind-current direction. What to make of that?

We spend hours chipping, prying, and breaking thin sandstone sheets away from the bedding planes. Late in the afternoon, to use the best shadows, we take another round of photos.

We walk to the base of the Nugget, then to the top — nothing. We find shotgun shells, beer and pop cans, abandoned shoes, broken glass, cold fires, wires, paper, an animal trap — artifacts of a locale too near the highway.

The orange sandstone inclines to the northeast, part of an immense hogback on the northeast flank of the Wind River Mountains. All the trails and burrows occur within a small strike-valley filled with junipers, pinyons, and low brush. The last light softens at sundown and shades the orange sandstone, the evergreens, and the valley. The magical dark light of sunset falls on a beetle crossing a sand dune, illuminating a scene that seems more Early Jurassic than Late Recent.

Sunday. We get up at seven o'clock, scrub ourselves, and put on our old field clothes. For breakfast we eat oatmeal, eggs, potatoes, and toast. The *Casper Star-Tribune* carries an entertaining melange of 1- and 2-day-old baseball scores, and distressingly complete accounts of NFL exhibition football games. We take all the paper to the car to wrap the samples we hope to find.

Boyd says we can, and we must, do three domes today — Lander, Dallas, Derby. All are old oil fields near Lander and show fine exposures of Nugget Sandstone. All are large sheepherder anticlines — up-folds that even a sheepherder could map, Doc Blackstone used to say.

We reached Lander Dome (sometimes called Hudson Dome) about 9:30. The air is hot and hazy from small fires in the oil field. Fires are frequent in oil fields. Today they are burning off a sump and a pile of junk.

The great anticline is sharp and is tightly folded, with the steeper flank on the west. Surface closure reaches 3000 feet. Producers & Refiners Corp. drilled the discovery well in 1909. Drilling continues. The productive rocks are carbonate beds and sandstone in the Phosphoria and Tensleep formations of Permian and Pennsylvanian age. Porosity is high, 15–20%. Wells are shallow, a few thousand feet. The oil is black, low in gravity, high in sulfur, and likely originated in the Phosphoria formation. More than 5 million barrels have been produced. Petroleum geologists who work in the Wyoming oil patch carry such summaries around in their heads — walking worlds of figures and geological observations.

The most complete Nugget outcrops are to the west. We decide to separate and make traverses perpendicular to the axis of the anticline. We review the characteristics of rock (eolian sandstone) that contains the trace fossils, ponder the job opportunities in geology for MS graduates (poor), discuss the ancient erosional surface at the base of the Nugget (well exposed here), and have a drink of water (warm).

Boyd is soon out of sight. He climbs quickly with sure step, displaying his great energy.

I take a less direct route across the upturned rocks, one that takes me through nettles, ants, and talus from the steep Nugget cliff. The lower part of the formation shows little of the characteristic trace-fossil rock. The upper part has the right rock, but only small expanses of bedding planes. No fossils.

Lander Dome is a failure. We spend 2 hours walking, crouching, and crawling over it, to find nothing. A few things look like trails; they may be trails, inorganic structures, or apparitions from bad dreams that I'm exorcising. Some sandstone and siltstone apparently is completely disturbed by organisms. There is no bedding in them.

Nothing is left but to eat fig newtons, drink warm water, and go on to Derby Dome. Boyd writes for several minutes. He is careful to record everything — failures and successes.

Derby Dome, another great anticline, is only a little way — 12 miles or so — southeast of Lander. Highway 287 passes through the field. It is only right that generations of Western kids have seen the grand structure expressed in red and orange Triassic and Jurassic rocks, then gone on to Laramie to study geology.

While Boyd gathers the gear he wants, I study the maps. Derby and Dallas anticlines lie on a line of folding markedly asymmetric to the west and southwest, extending northwest for about 50 miles parallel and northeast of the Wind River Mountains. In early reports, this folding is the Shoshone anticline whose long sinuous fold axis bends, and is broken across the Wind River Basin, never getting much more than 6 miles from the mountains. In some places, anticlines are thrust out over the syncline separating the fold from the Wind Rivers. Porosity in the Phosphoria and Tensleep formations is high, but oil production at Derby Dome has been small.

Our traverses are close together, within shouting distance if the light breeze goes down and Boyd's bad ear isn't turned toward me. He lost hearing in the ear quickly and unexpectedly on a trip to New York to study pelecypods. "A virus or something" is the way the doctors explained it. He moves the good ear like an antenna to gather in what he can.

The eolian sandstone appears barren. Enigmatic structures, nothing else. Boyd finds nothing to the south in the sandstone.

Someone has tied red cloth strips to sandstone knobs and to bushes. We conclude that they mark a celebration site — homage to the Great God Orange Quartz or the Feast of Black Oil Fruition. Possibly they mark a survey route. They don't mark fossil tracings.

Boyd takes a sharply incised gully, I take the one to the south of him. The V-shaped ravines are narrow and close together.

At the toe of the ravine there is hardly room to walk along the channel bottom. I straddle the channel until it widens. The last flood swept the bottom clean and uncovered bare rock extending nearly to the head of the ravine. Climbing slowly, I meander back and forth across the channel to cover the most ground.

Vertical burrows jolt my daydream. Like small buttons in a jar dumped on the floor, they dot the carbonate bed, falling singly here, there abundantly. The rock looks like a lake deposit, certainly not an eolian deposit. Wave ripple marks, the carbonate beds, and small shrinkage cracks suggest the origin. A new find in different rock. I climb out of the ravine and see Boyd at the crest of the hill.

He is elated. He pulls out a hand lens and spreads his bony body across the carbonate. He pries a piece away to make sure the burrows pass through the bed and are not just inorganic surface features. Within a few feet are hundreds of excavations that were made in unconsolidated sediment. We measure 60: burrow diameter, 0.3 inches, with an internal diameter in a third of them of 0.1 inches. The bump-like structures are raised a few millimeters in relief.

We collect burrows and rocks for thin-section study. By the time we wrap the rock, the world's news — wars, politicians, crime, criminals — is gone and we sacrifice the sports pages to wrapping. A 75¢ Sunday paper doesn't go far.

We take photos and make last-minute notes about the rocks and burrows. Boyd spots ripple marks in beds above us — current types. He wishes he had a long-sleeve shirt on, he says. His arms are the orange-pink of some siltstone and sandstone interbedded with the carbonate.

"We must give this discovery to the world ungrudgingly," I say.

Boyd grins. "The world is not holding its breath."

We head northwest toward Dallas Dome, the first field discovered in Wyoming. The 1000 feet of surface closure, the location, and an oil seep made the dome an obvious choice to explore. Mike Murphy, gold prospector, and one Dr. Graf of Omaha drilled successfully in 1884 a few feet from an oil seep Washington Irving described in 1837 in *The Adventures of Captain Bonneville*. They found oil at 300 feet in fractured red siltstone of the Chugwater Formation, a Triassic deposit. The field was small. During the early 1900s other operators discovered oil at a few thousand feet in the Phosphoria in dolomite and in the Tensleep in sandstone. More than 5 million barrels of the low-gravity, high-sulfur, black oil have been produced.

The oil in the Chugwater siltstone came from the older reservoirs, and the oil in them had its source in the Phosphoria Formation. The oil in the Phosphoria came from marine microörganisms in the ancient sea, and the ancient sea came from the west.

I don't say any of this, and Boyd doesn't crown me with a geologic hammer. We separate to look at the Nugget Sandstone on Dallas Dome. He takes the steep cliffs rising from the Little Popo Agie (pronounced Popo 'zha) River. I follow the dirt road and the river downstream. At the first large draw, I leave the road and cut obliquely up into the formation. The lower beds are barren. Above them, I find a carbonate bed similar to the one we saw at Derby Dome, one that contains some vertical burrows.

I climb higher into the cross-stratified orange sandstone that produces vast amounts of petroleum in the thrust belt west of us, enough petroleum there and on other Wyoming state lands to allow low property taxes and no state income tax.

The eolian sandstone is present and bedding planes are exposed, but ancient trails are not evident. The promising sequence at the top proves lifeless. I traverse down to the middle of the sandstone and back again to the top.

In the wide draw to the east a doe crashes through the brush and bursts out onto the orange sandstone ridge across the valley. She stops, sniffs, and runs higher into the trees. I sit quietly, willing her to stop, to come back. She walks from a grove, turns, and continues downslope toward the draw she came from. Her fawn is there, I think, and if I don't move or blink she may come back to the fawn. She walks a hundred feet toward the draw. When she stops, I hold my breath. But nothing can hold her. She turns and runs lightly back across the sandstone ledges to the grove and disappears among the trees. I wait 10 minutes, but she never appears. Unless Boyd saw her, he'll say this is just another of my deer stories.

The burrowed carbonate bed runs down to the draw, making a V that would point upstream if a stream ran down the dry draw. High above the draw hawks wheel and trace in flight smaller and smaller ellipses until they break off and rise quickly to circle and swoop again. The sun sinks low. The long curve of the Wind Rivers disappears in the fiery sky. Beside the river the air is cool and evening sounds break out. A pumping well squeaks harshly. The pump needs oiling.

Boyd says he saw the carbonate bed on his traverse. No burrows. He describes an odd imprint in red siltstone and pulls out the rock. The siltstone takes the last of the *Star-Tribune*. Boyd figures that the quality of the paper is dangerously close to inadequate for wrapping trace fossils or anything else. We lean against the Subaru, enjoying the magical light and the sounds of evening. The three domes and all the collecting went fast, too fast. I ask Boyd about the future of the Subaru, which has developed an expensive-sounding deep growl. He says he's in the market. He recalls Doc Knight, former head of the geology department at Wyoming, extolling the virtues of the Franklin. Doc often said, with little prompting, that you could replace the drive shaft in a Franklin in the field, implying that he carried spare drive shafts and whatever else was needed to keep the Franklin going. A long jump to electronic systems, I say.

As Boyd gets in the car I put my camera on the back seat and take a last look toward the sandstone above the draw. The search has been quietly exciting in a way that teaching 300 students physical geology is not. Following the trails and burrows led us to lakes and deserts and to clear hot days at ancient assemblies. We don't know what kind of creatures they were. Just what the tracemakers were doing, when they arose, and how long they survived is also unknown. Yet, to know something, however little, of their world makes me feel better in my world.

A Mine Dies

On a recent Friday — April 9, 1982 — a metal mine died. That wasn't unusual; mines having been closing lately all over the world.

Three of us went to see Noranda's Ontario project in the Park City district in the Wasatch Mountains, east of Salt Lake City. I awoke at 5:30 A.M. — without the nerve-shattering sound of an alarm — and splashed cold water on my face. Before Jim and Frank came, I had a breakfast of grape juice, Shredded Wheat, and tea — a fitting fare for a middle-aged professor with acidic stomach. When my colleagues drove up, I was in the back yard, where I'd noticed that spring was later than at Walden Pond.

We were in Park City before 7 A.M. In the late 1800s, it was one of the liveliest boomtowns in the West. Millions of dollars were made and lost in speculations. Cornishmen, Scotsmen, and Irishmen came to Park City to work the mines. Scandinavian lumbermen ran the sawmills. Chinese laborers, laid off by the railroad, opened laundries, served as waiters and porters, and sold vegetables.

Park City grew toward the Sun in high thin buildings. The narrow valley prohibited the grand sprawl of a Salt Lake City. Later on, through the first half of the 20th century, Park City's fortunes rose and fell with the general economy. Mining was nearly continuous, but of minor economic importance during the

1960s. The town turned instead to real-estate development, ski resorts, and tourist attractions. Condominiums, hotels, restaurants, ski runs, golf courses, tennis courts, swimming pools — necessities for the well-to-do.

We drove up the narrow, main street. Uncertain of the mine's location, we took three wrong turns south of town. Still lost, we drove beyond the town water tank to a dead-end of a 15-foot snow bank. By 7:30 we had found the mine. Al Gordon, a serene, auburn-bearded engineer, said that all but six persons would be let go before the day was over — a gentler term than fired. We waited half an hour for two magazine editors who said they wanted to go underground. They didn't come.

The hoist supposedly holds 12. It's close quarters with four. We descend toward the 2200-foot level. Part of the shaft is new, part is old. Rock and wire and dark openings pass slowly. The minutes are interminable; I can't recall ever sinking so deliberately. The air is cool and musty. I smell wet rock and hear running water. Will the hoist man — hoist woman, I sense — be there to bring us back up, or will she be let go before we finish? My stomach is uneasy and my nose itches. I look intensely at the rocks all the way to the 2200-foot level. Al and Jim Whelan, a gray-bearded, pipe-smoking, geology professor, joke as we make our way in rubber boots and yellow wetsuits. We stay on a wooden walkway between the rail tracks. Water, a few inches to a foot and a half deep, flows beside the tracks, forming ripple marks in sediment on the tunnel floor. Ripples are unexpected, deep within a mountain.

We walk a long way. The distance between Al in front and Frank Beales and Jim is less than between them and me. The breeze is cool, but warmer than on the mountain top. "It's a lot warmer here than in Toronto," Frank says. Frank is of medium height, and bald with deep wrinkles at the corners of his eyes, a thin gray mustache, and a quick wide smile. Among us, he is the specialist in these rocks.

Ore cars, devoid of their engines, sit abandoned all along the tracks. The water pumps still work, will run their last few strokes, and soon the cars and the mine will fill with water — 1500 feet

— a lake inside a mountain greater than all the swimming pools in Park City. The miners, some with ancestors in the Ontario mine a hundred years ago, cannot afford the new Park City. They reside instead in Heber City, a middle-class town east of the Wasatch Range.

I catch up. Al is looking up into the dark. He climbs to check for carbon monoxide. The ladder looks narrow and rusty, its height alarming. The three of them climb above the tunnel out of sight into a high shaft. My turn is next. I'll wait here, I decide, until they return. I pull my helmet down, surprised that the light is shining directly on the foot of the ladder. The first six rungs go quickly. My knees are weak. The rubber bib overalls and clumsy boots and 2-pound hammer hold me down. I climb higher. The hammer is in my right hand, clenched tightly against each new rung. "Dammit." I thump my head. The hard hat falls forward, the light swings free. I catch the lamp in the darkness with my left hand. I'm close against the ladder, but can back down in a minute. I won't continue. The hell with it. I reach up and put the hat back on; Frank reaches down to take my hammer. He's crouching on a wet 9-inch plank. The plank on my side of the opening is the same width. I let go, fearfully, pulling myself onto the timber. I inch along the plank to a tiny platform above the lower tunnel. Frank's board is slippery, but only a few steps from the rock face. It glistens with lead and zinc ore, metallic and resinous from the lusters of the galena and sphalerite, and wet from water dripping from the roof. I pick up a brassy yellow cube of pyrite — fool's gold.

I hear Al say, "We need 7 ounces of silver and 4–5% lead and zinc. We haven't got it. The first ore here assayed 96 ounces a ton. That is ancient history."

"What's your investment?" Jim asks.

"Thirty million on the lease."

"You gave it a good shot."

"The veins are narrow. Not rich enough in silver."

"Metal prices haven't been this low for a long time," says Jim.

I wait until Frank is clear below. Going down the ladder is easier than going up. My legs are strong. I feel fine. I start a small rock fall. No harm done.

One at a time we climb another ladder deeper in the mine. So little space. I pull myself up onto the planks. Crawling on my hands and knees across the timbers, I feel like a boy crossing a pipe over a canal he's not supposed to be near. The wood is wet and muddy.

Al says: "We couldn't have got the editors up here. Just as well they didn't come."

He is happy they aren't along, wanting this last trip to be to every nook of the mine he knows.

"You barely got one of the professors up here," I reply.

The rocks dip 30° to the right. The vein — more than a foot wide — shines in the light of our lamps. The brecciated zone below the vein contains pieces of limestone the size of grapefruit. Wall rock is gray and brown. The hanging wall has moved over the fault plane above the foot wall. No! Hanging wall and foot wall have moved together. The cut is too small to reveal how much movement has occurred. I see an empty Mountain Dew can, flattened and silvery among the gangue. The debris of the careless miner's great-grandfather may be buried somewhere below the rubble, along a tunnel out from #3 shaft.

A dark tunnel leads into a black hole above us on the left. As Frank studies the vein and the breccia, I stare up the incline to the hole. The exposures are superior here; we won't need to go in. Frank and I differ on the geometry of the rocks. He is thoughtful — his approach to mineral deposits. He crawls out onto a ledge. I don't. He picks up a rusty ax; none of us wants to carry it back. He throws it onto the ledge. The sounds of the mine — water running, water dripping and splashing, dust and pebbles falling, air moving along faults and fractures and tunnels — all reach us in this new shaft. The mountain groans and sighs and heaves in cries undetectable above ground. We sense the miners who have worked here — at one time 200 a day — and the ghosts of those who died underground. Our voices — perhaps the last ever to speak in the shaft — are muffled between long rock walls. Frank wants to try to go on into the black shaft. "Sure," Al says, "we can do it."

The incline steepens as we climb. This side tunnel is without streams, but the rocks are wet and slippery. I grab a rope along the wall and hope the end is anchored. I drop the rope and grab a pipe; its presence here is unusual, less certain by far than seeing the Mountain Dew can. Jim slips and falls but recovers. He is agile and the right size for mines. Galena and pyrite sparkle in the light as I turn toward the open face. We carry the light with us to the top. Black is becoming gray. Frank moves as close to the vein as he can, collecting samples of the most mineralized rock. I break a piece away from its host and hand it to Jim. He looks for the brilliant crystals of tetrahedrite, which he believes hold the silver. He drops the ore.

Faults and veins trend east–west in the Park City district. The anticline trends north–south. Al is above us, as high as he will let us go. He wants to move back down the slope and on into the main part of the mine. There are other rocks and works he plans to see. I agree. Geologists putter along; engineers move rock. Frank is satisfied at last. He yields gracefully on the structure of the vein and fault.

Pools of water are deeper in the oldest part of the mine. Walking beside the rails, I slip. The water is above my knees. Sounds of an underground river flowing along a drain tunnel reach us. Jim falls with a shout into a hollow, and bounces up sputtering. There are rock falls across the floor; some look new. The mine is dark and old, long ago returned — its youth spent — to the mountain. The rock walls are older, less sheer than elsewhere. Their strength holds them above the floor of the shaft. Timbers and pipe lie deserted in heaps. The turbulent stream in the drain tunnel is close by. I climb through a mound of pipes and timbers and abandoned tools, following Al to the lean vein.

"They didn't leave much of a pillar," he says, pointing to pinched rock holding the tunnel roof above us. Al sits down on a rock bench below the vein. "They didn't know how to cut them for safety."

Piles of fallen pebbles cover the floor of the gallery behind Al. We are near the major strike of 1872, made in Ontario Canyon, just east of the Ontario #3 shaft. Samples held 100–400 ounces of silver per ton. Those claims sold for $27,000, to become the foundation of a company that eventually paid $15 million in dividends through 1956. I recall the narrow lenses we've seen and Noranda's investment in them.

We walk toward #6 shaft. Six is a lucky number, my old basketball number. But I'm falling farther behind than before. They
go around a corner. I can barely hear them. The shaft is dark
without their lights. The mine is still except for the stream. I'm
more than a mile above sea level sheltered deep within a vertebra
of the Rocky Mountains. Is anyone sheltered? There are no shel-
ters from nuclear bombs. The mine is a shelter for hours, not
years. Men struggled over 10 decades to dig the mine; in days
the tunnels will hold a lake. "Dammit." I bump my head again.
Is my neck damaged forever? Middle age. People don't fit easily
into mines. Basketball players don't visit mines. To work, miners
must crouch and squat and crawl. I hurry to catch up.

The light ahead is brighter than our lamps give. The platform
lights. I'll be there in 30 steps.

Al pulls the rope, ringing for the hoist. No answer. He phones.
His voice is soft and calm. He talks to someone who will have
us picked up. Will the cage come?

Bland graffiti: "Don't Be Fooled! Beetle Gray Is a Dork! (And
Then Some.)"

Above Jim's head is: "Dubie Sleeps on the Job." Then, "What
About Buzzard Lips?"

Jim asks Al where he lives.

"Midvale," he says.

Al likes Midvale.

I hear the hoist. I feel relief and see it in the men's faces. Al
bangs the gate and pulls the signal rope. We ride up the #6 shaft
to the 1500-foot level. We walk again. We walk some more. It
is our longest tramp. We're crossing to #3. Al, Frank, Jim, Duke.
The order of our column stays the same. At field camp, the
summer of '49, I was the first geologist to reach the top of the
granite peaks. On some mornings, that was the most important
thing. But that was '49.

It is fine to be underground today. I'm tagging behind our
group, but I'm ahead of the world. The companionship of rocks
and solitude. The Ontario mine was a bonanza, a bust, and a big
but unprofitable mine and, finally, it is broke again.

I walk slower. This is the darkest tunnel. Dust is in the stale
air. They've stopped. Why are they huddled together? Two of

them are standing over the third in the middle of the tracks. Frank is pulling on his leg; Jim is tugging at Frank's ankle. Al is hitting Frank's toe with a hammer, which bounces off the steel-plate. Now I'm alongside Frank's right foot is caught between the rail and the plank, a professorial leg anchored to an abandoned track and mine.

"Lucky there isn't a string of cars coming," Jim says.

"I've never seen this happen before," Al says.

"We'll have to shoot Frank," I say.

Frank grunts and squirms. He is upset but unties the boot and slips his foot out. The boot remains fast.

"Can you turn the boot sideways, Frank?" I ask.

"What do you think I'm doing?"

Frank is embarrassed. We stand beside him, not knowing quite what to do. Al and Frank give the boot one more tremendous tug, breaking it free. We continue to #3 shaft.

The ride to the top is faster, it seems, than the ride to the bottom. I step out after Jim, relieved and delighted to be standing on the mountain. The workings are below us — silent, gray, dusty, wet. Standing in the modern, mustard-colored building, the miner's clothes around us, I fancy I hear, for a moment, a room of voices — men dressing for the next shift. We wash up and walk outside where the sun is high and bright in a cloudless sky. A chilly breeze blows across the snow straight down the valley. My eyes water and hurt. I close them and point myself toward the truck. As I reach the truck, Frank and Jim are eating doughnuts and drinking black coffee from a thermos. I put on sunglasses. The new Park City is below us, unaware that almost the last part of the old Park City may have died today. I eat an orange and drink the last cup of coffee, one of my five a year. We start down the road, the still mine behind us, the new Park City ahead.

The Swimmer

We have never entered into an animal's mind and we cannot know what it is like, or even if it exists. The risk of attributing too much is no greater than the risk of attributing too little.

— Joseph Wood Krutch

Orange-bedecked men dotted the hills, and the sounds of gunfire reached us off and on all morning. Every few minutes we saw a truck carrying two or three hunters race down the road, followed soon by more gun shots.

It was the last day of the hunting season. The first weekend the roads had been wet and in many places they were slippery. The swelling roadbed clays collected rain and snow and expanded. Hunters were stuck everywhere on roads far below the better hunting ground of the high plateaus and canyons. In the hunting camps, the men talked and ate and drank and two of them argued strongly enough that one shot the other in the face, killing him as the season started. Later on, a young man drowned. Not far away, the emergency brake of an old man's pickup failed on a hill and the truck ran over him, breaking his legs. A blood clot formed and he died. A middle-aged woman suffered a heart attack. Her husband slid off the county road on his way for help and got stuck in mud eroded from 80-million-year-old Cretaceous claystone. His wife's heart stopped. Newspapers reported this to be a normal hunting season, even uneventful, compared with the large number of deer and people killed in previous years.

None of us saw the deer run over the top of the dam or enter the water. But within a few minutes everyone in our field party was aware that a large deer had come from the fields below the dam, frightened, and was swimming toward the center of the lake. Several marveled at how quickly she must have picked her way through the sandstone-strewn, boulder face of the dam down to the lake edge. Not everyone knew deer could swim. From where we stood, on the extreme north end of the dam, not everyone could see that it was a large doe and not a buck.

There were four geologists in our field party. We had been looking at beetle trails on sand dune slopes in ancient wind-blown strata, sandstone that was also used to face the dam. We found several trackways where the paths of beetles had crossed some 180 million years ago. When we saw the doe, we were on the way back from looking at Cretaceous beach deposits in pale brown and yellowish-brown sandstone.

The doe swam confidently toward the center of the lake, her head still, the ridge of her back breaking the brown water. Her long tail, white on the underside, streamed out behind her as she met the current obliquely. The white band across her nose and the white patch on her throat were tiny banners riding olive-gray ripple crests as a breeze came out of the southwest.

She veered slightly with the wind, no longer swimming parallel with the length of the lake. The breeze shifted and blew straight down the lake. She continued northeast. Several fishermen on the south end of the dam sat close together, a few feet above muddy water littered with paper cartons, a truck tire, and the ubiquitous beer cans of Utah lakes and foothills. No clouds. The sky was dark blue from horizon to horizon. The strange light the sun made in the water hid the doe's head for a moment. She swam steadily on a line that would take her into a bay half a mile up the north shore.

The doe moved through a small patch of driftwood on the lake surface, neither changing course nor seeming to notice the wood and broken brush across her path. My younger daughter swam like that in high-school swimming meets — slowly, strongly and steadily for long distances.

We walked 30 yards south. The others were talking. I was not listening. I was trying to keep the doe in sight. I did not think I had ever seen a finer-looking deer. I thought she belonged with a small bunch I had seen resting in the aspens on the granite or grazing in a small meadow among blue spruce.

I imagined that hunters shooting below the dam had frightened other does away. North of the lake across the floodplain were low foothills. The does had run into the foothills above the fields, I thought.

The sun had taken the shadows from below the dam. It was clear and bright and warm, presaging good weather for several days. The air barely rippled the lake's surface.

Each of us had a different idea of what had driven the doe into the lake. All of us were short of facts. We had been reconstructing what had happened on a Cretaceous beach 80 million years ago, but were uncertain of what happened 20 minutes before on a Saturday morning in late October.

We could still see the doe. A duck paddled within 20 feet of her, being careful to come no closer. The deer obliviously swam on, head high and steady in the water. Nothing in the lake could harm her. Surely no hunter would be so depraved as to shoot a swimming deer far from shore, said someone.

She had swum a long way. The gently undulating surface rose to cover her from time to time, rising and falling rhythmically but almost imperceptibly as the distance between us grew. She moved slowly on, still in the direction of the narrow bay on the north shore. I thanked my mother and father for the far-sighted eyes of a field man. The lake was very dark and the light formed dancing beacons in the distance. No one said anything. The others strained to see the swimmer.

I lost sight of her. I imagined she had vanished back into the Miocene when deer were smaller and species fewer. By the end of the Pliocene, the types were numerous, many surviving to modern times. Paleontologists have given them a lot of names, assigning the swimmer to *Odocoileus virginianus* — white-tailed deer of the Americas.

The breeze died and the water was smooth. I could just barely see the deer swimming in close to the bank at the mouth of the bay. The shale and sandstone fell off steeply and it looked as if she couldn't get out. I lost sight of her as she swam on into the shallow end.

We walked across the dam to the road. A truck went by. The hunters inside studied the low, brush-covered hills. The fishermen said they hadn't seen the deer at all. One of them pulled a 3-pound sucker onto the bank and beat its head against an orange sandstone boulder. Even so, it was a fine morning. The sun and distant water were very bright again and the air was clear. For some reason, I felt sure the swimmer would make it through the hunting season.

Time in the Field

On a Saturday morning at five o'clock, I got up and shook Dane's 17-year-old shoulder. He wasn't willing to move or talk. If left alone, he would sleep till noon. I didn't intend to be that indulgent but decided to give him an hour more.

I had breakfast in our room at the Green Acres Lodge in Heber City and looked at geologic maps of northeast Utah. By then, the sun was up and it was time to give him a no-nonsense shake. He rolled over and said something. The words were incoherent or were sounds of his generation I didn't recognize. It was encouraging that he had moved. I then took half an hour to study glauconitic sandstone — the ticking of the glauconite clocks among the quartz crystals of the sandstone could not be heard above Dane's breathing, but the sandstone was critical to my study. He stirred again. I hurried to prop him up before he stopped moving. He groaned at being caught half awake, but opened his eyes and spoke. "Is there time for breakfast? Where are we going and what are we going to do?" he asked. For someone still in a daze, his questions were remarkably organized and appropriate. When he had dressed and watered a sheaf of brown wavy hair, we went across the street to the Wagon Wheel Cafe. Afterwards, he was more alert and I asked him if he would like to drive. Stretching to his full 6 feet, and yawning, he said, "Not particularly."

Teenagers sometimes speak that way to their parents. Parents, they believe, inhabit a different world, and they are right. But later that day we talked about rocks, hawks, picket pens, and beavers. It was a fine long Saturday — men and animals, rocks and time.

From the beginning of the day to the end, the element of time was a central issue, beginning with getting my son to join the day. The rocks brought time to mind in a different context — as evidence of periods of time beyond imagining and the only record of hundreds of millions of years. But rocks surrender history grudgingly. Their deposition is discontinuous. There is no record of what happened for days, for decades, for hundreds of thousands — even millions — of years. Time gaps are longer than rock records. Geologic time and the earth's history stretch incomprehensibly to nebula and fly to the sun. For Dane, time ambled; for me, it was a juggernaut. Paradoxes of time — relative and absolute.

The sun behind us, we started for Provo. I drove, Dane dozed. In an hour and a half, we left U.S. 50 southeast of Provo and turned on to one of those old dirt trails whose map symbol indicates you are embarking on an adventure of high risk. According to the legend, the road was unimproved; its narrowness and steepness would save it from becoming improved for years. I was grateful for that, but to drive on any road in an elderly Fiat that constantly needs repairing is to enter an uncertain future. A motorist once shouted, while we were working on the engine, "Fix it again tomorrow."

It seemed to me, as we moved farther from civilization, that we became more civilized. Dane, now alert and energetic, helped with the measurements. It was warm working in the sun and I could smell him and he me. We were almost alone in a strange country. The rust of our lives began to fall away. There was good humor in being out there and alive and moving about. We listened to the animals, the river, and the soft wind. The day became timeless — a day of geology among mountains whose folded strata took aeons to reach their complicated structure.

We worked among the rocks of the Wasatch Range in north-central Utah and along the Diamond Fork River, passing between towering conglomerate walls of the Price River Formation. Diamond Fork flows to the southwest; the ancient stream that carried the clasts of the conglomerate flowed to the east, a torrent of pebbles and boulders in a roiling flood from rising highlands on the west. We determined current directions preserved in the storm's debris and then stopped to eat along Diamond Fork in the shadow of quartzite and limestone boulders — boulders that measure the hours of the ancient flood. Dane drank three Dr. Peppers and ate a can of sardines, a can of Vienna sausages, two handfuls of crackers, two peaches, and half a package of cream cheese. It had been 5 hours since the six pancakes, ham, and hot chocolate he had for breakfast. In the summer, his days, like those of other young men his age, revolve around eating, sleeping, and being with friends.

The road climbed higher across the western flank of the Uinta Basin into sedimentary beds of simple structure. West of us, on the skyline, we could see cirque basins of the Wasatch Range that held snow along the crest. Snow would remain in the highest of them through the summer. All of the basins would be filled during the fall and winter.

I asked Dane if he knew the name of the layers we were driving on. After a long pause, he guessed and said, "Wasatch." I was pleased. "Close," I said. "We're in the Green River Formation that overlies the Wasatch."

"I should have known," he said. Beginning when he was a little boy, striding around the field in his first cowboys boots, we'd look each summer at Green River rocks. The formation had become an obsession.

Not far east of us, we'd seen fossil tracks of sandpipers, kildeers, and gulls crossing rocks of carbonate mud flats marginal to the ancient Green River lake. Three-toed mammals came down to drink along the same shore — a shore that we five-toed mammals had also stood on to record the geology. Lizards, their feet scurrying this way and that over the feeding grounds, must have frightened the ancient birds. Frequent rainstorms freshened the water, and impressions of the vanished raindrops were left to dot the surfaces marked by bird tracks. The tracks and raindrop impressions on the bedding planes recorded only a moment, perhaps a sunny late afternoon in August interrupted by a thunder shower and the flooding out across the flat of streams flowing into the lake. For an instant, I could see the animals of the Green River lake shore come alive again — wheeling bands of birds, fish and reptiles, mammals, and insects. Their calls and cries filled the air above the road. The smell of ancient drying mud flats replaced the dust in my nose, and my allergies disappeared for the first time in weeks. Before I was ready, the reverie receded to my inner brain. Afterwards, as we rode through scrub oak and aspen, Dane seemed to become me and I became my father. In those seconds, the generations passed each other in a rush, leaving a chill.

Our silence made the noises of the Fiat seem louder — the whine of the tiny motor, squeaky brakes squalling on the sharp turns, and the thump of wornout shock absorbers.

I asked Dane how he'd like to be a geologist. He grinned but didn't answer. He'd heard the question many times; the words were a ritual between us, a joke shared on good days.

"What do you want to be?" I asked.

"An engineer," he said. "Electrical or mechanical."

As I recall, his last answer to that question was that he wanted to be a photographer with a background in chemistry who builds racing cars on the side.

I told him I thought petroleum engineering was exciting, not wishing to leave completely the idea that he might discover geology for himself even if it was through petroleum engineering. He didn't say anything. A day or so later he asked about petroleum engineering.

As we drove up a narrow valley, he pointed out a beaver dam. I looked to the east and saw the dam below us. He proceeded to tell me about beavers. He said they are the engineers among animals and can build dams hundreds of feet wide across valleys. Beavers build two entrances to their lodge underwater, with the openings about 3 feet below the surface. Sometimes they build two stories, he said. If there are only a few beavers left in a valley, they may lose their ability to build a house.

But this was not the case here. Their architectural achievements filled the valleys. When I asked him if he thought their skills are learned or instinctive, he said beavers learn and pass their knowledge on to children and neighbors. "They mate for life," he added.

I decided he knows more about beavers than I do and about a lot of other things as well.

Clouds had formed southwest of us and were moving across the Wasatch Mountains toward the Uinta Basin. It was early in the afternoon before I noticed them. The distant peaks of the Wasatch Range had become small and undistinguishable. From our position, we couldn't tell if far-off mountains had turned into clouds, or dark clouds had evolved into limestone peaks. It began to drizzle as we stopped at the summit on Strawberry Ridge. We got out and looked out across the country where we'd been. A light breeze blew over the ridge from the southwest. The rain was cool on our faces, but wasn't enough to bind the dust still rising from the road below.

"She's diving for her lunch," Dane said.

"What?" I asked, lost in my own thoughts.

"The hawk's diving for lunch," he said. "I think it's a female. They're bigger."

"Did you get the dive?" I asked.

"No," Dane said, continuing to adjust his camera. "She was too fast. Just folded her wings and dropped."

We walked down a steep bank and through low brush in the direction of the hawk's fall. Wild rose bushes cast shadows on bluish-purple larkspurs and baby-pink sticky geraniums. The incarnadine terminal umbel of a shootingstar on a high flowering stalk flamed in the shade below an aspen. The mountain wild flowers were blooming in their short season.

"She may come back up," I said.

"I hope so," Dane said.

He took pictures of all the mountains south of us. The narrow valleys through the trees were strings carelessly dropped on the land from a great height. The mountains looked huge, silhouetted against the blue-gray sky. The clouds cleared momentarily.

"Here she comes," Dane said. "There's two of 'em. Her mate's with her."

The hawks rose quickly at a high angle to the slope. They turned and glided toward the aspens behind us, two flyers alone, still hunting.

"They're coming over," Dane said. "I can get 'em now."

His head and the camera moved with the hawks as he made adjustments for their flight. I was sure they would pass over us. Suddenly, as if alarmed, they changed direction and rose higher.

We watched the hawks move out of sight. Dane was angry but didn't say anything until he'd put the lens cap on his camera.

"For Chrissake," he said. "It's the cooling motor on the Fiat. They heard it."

We turned and walked toward the car on the ridge above us. A field of sunflowers in bloom stretched across a clearing into the aspens and down the grade through a pass toward Strawberry Reservoir. In a week, the flowers would be gone and the clearing would be bare for another year. While watching the hawks, we'd forgotten the lake. For a moment, I felt giddy — a flash of middle-aged dizziness — and the yellow field and the white barks of the aspens and their green leaves whirled concentrically together into one of Van Gogh's wild visions near Arles. We moved down among the sunflowers.

A jeep passed on the road; its single passenger looked lonely. I watched him disappear into the aspens before turning back to Dane, who was taking pictures of sunflowers and the lake. The hawks had not returned. It had stopped drizzling and the sun was bright again. The calm was broken by a single rifle shot. The sound came from nearby. An image of the driver stopping and shooting himself would not go away. I knew I'd have to see, but Dane should stay. If the man has really killed himself, why did he have to do it now? These crazy scenes and feelings passed through my mind, a mind not entirely recovered from the fierceness of spring politics in academe.

My fears for the man and the day were calmed by the reality of a truck on the road and the appearance of two shooters. They rode toward us in the back of a pick-up, standing and leaning on a blanket spread across the top. Their rifles moved in arcs across everything in front of them and swung to cover their rear which was unthreatened. The two closely resembled young men in Heber City who patrol Main Street at night, staring hard at strangers. At first they examined us as they would a deer and then suspiciously, as they would a rival for their girl friends. Their truck went over the summit, however, before they fired again.

"They're shooting picket pens," Dane said.

"They shouldn't be shooting anything. Nothing's in season."

"Fools."

We walked back to the car.

"Did you get all the pictures you wanted?" I asked.

"Yeah," Dane said. "Except the hawks."

We drove on. The lake was still after the rain. "It would be a good afternoon for water skiing," Dane said. "Smooth as glass. No wind."

The lake was a large one, and winds could make it wild, but it was usually tame. There were fishing shacks and trailers sprinkled around the shore. The buildings were in disrepair and some of them looked unlikely to make it through a hard winter. The aspens were rooted well above the lake; the shore was barren and rocky with low brush and sparse grass. Rock and soil were red. The lake had once been a famous fishing place, and we had fished there, but algae were taking over, and for several years the big trout had been caught in other lakes. There weren't many places around the lake that seemed remote anymore. It would be hard for a young man to get lost. The country was wilder on Strawberry Ridge and, from a distance, the lake and its shore sprawled romantically, a scene in the Rocky Mountains that Constable might have painted.

U.S. 40, the road to Heber City through Daniels Canyon, came up quickly and we turned onto it. The smell of the light rain still filled the canyon, giving the air a fragrance and cleanness undiminished by vehicle exhausts. Saturday shoppers were in Heber City; Saturday night revelers had not begun their runs from ranches in the basin to the lights of Main Street. Dane's head fell forward several times before he settled back against the seat and slept. In jeans and a brown shirt with the sleeves rolled to the elbows, he looked like a young field geologist resting for the next day's work. His body, for an instant, was again mine, riding along in the evening with the senior geologist to Gillette in northern Wyoming. On that first job, we mapped an anticline in gray Cretaceous beds. For 3 weeks, we traced coal seams and siltstone through the sagebrush across low stubborn hills and down dry gentle valleys until a great anticline magically appeared on the plane table sheet. Had three decades passed? The smell of coal dust arose and was sweeter than the smell of raindrops on the evergreens in Daniels Canyon.

We drove slowly through the canyon along the flank of still another anticline. Then we crossed the Charleston thrust fault and coasted over the alluvial flat to Heber City. Long dark wands moved out from the mountain front on the west as the sun disappeared behind the mountain's black profile. The traverse in the field had been a sort of circle, the kind of ring children in kindergarten draw when the teacher asks for a triangle. The morning began with a sleeping son; he slept as we turned into the motel driveway. The day was a free one and wouldn't count on his years; I felt it hadn't been added to mine.

On the Road

Down the Durance in France

On Sunday if you had been with us, you would have seen a red vineyard,
all red like red wine. In the distance it turned to yellow, and then a green
sky with the sun, the earth after the rain violet, sparkling yellow here and
there where it caught the reflection of the setting sun.

— Vincent van Gogh, Arles,
November 1888, letter to Theo (The Red Vines)

The river of my boyhood flowed north undammed, flooding
in the spring over Mexican Town, soaking the bottom half of a
sack of my mother's beans and, checked in its flow over the
floodplain, depositing silt between the swelling beans.

When I was 10 I traced the river to the north in an atlas, past
Basin, Graybull, Lovell, into Montana. It turned northeast in the
direction of Forsyth. With the Porcupine and the Rosebud, it
joined the Yellowstone. The Yellowstone ran northeast to Glen-
dive, on to Sidney, carrying with it all the smaller rivers to the
white blank edge. The next two pages were missing, torn from
the book by someone in another class. The ocean was pale blue,
New Orleans was scarlet. The missing pages were green and
brown, I supposed.

The river turns east and then south and joins the Mississippi,
I learned in sixth grade. Rivers run into oceans. I made such
categorical pronouncements in those days.

The river of my boyhood was the Big Horn. I grew up on the
river, never far from it in my school years. My town was Worland,
which straddles the river on the fertile floodplain.

In my middle age — how much longer can I say middle age?
— I still gather sand from rivers on their way to oceans and sand
from beaches at land's edge. I am a collector and cataloguer of
grains eroded from ancient lands.

The Durance rises close to the Italian border in the High Alps near Briançon, pours south through the wide mountain gap it has cut, turns west below Cadarache, and runs northwest into the Rhône, joining it 3 miles southwest of Avignon. The Durance was once the scourge of Provence, but its 201-mile, irregular course has been tamed to serve the local economy. A network of canals carry water for irrigation, town use, and electricity.

The Rhône runs from the High Alps on the Swiss border to its wide delta, marked by the Berre Lagoon and the Camargue on the Mediterranean. I went with Earle McBride, a sand collector from Austin, Texas, to France in June to sample the sand of the Durance and the lower Rhône. We'd cut across country from Paris to the headwaters.

The Durance is braided below Briançon, and pebbly. Gravel caps lozenge-shaped longitudinal bars which become finer-grained downstream. Within the bars, the gravel is crudely bedded in horizontal layers, an upward fining of gravel and sand. We took a sample from one of the longitudinal bars near the river bank.

In the late afternoon we drove on, down the Durance, passing through very small villages, stopping to look at the river and sample the sand every few miles. Above Embrun we picked our way near the river edge over water-worn pebbles, through algae covered and aligned downstream, and through a shallow pool of stagnant yellow water with leaves floating on the surface. The mountains to the north were gray and green. The wide valley was a glacial trough, rounded in profile to the valley floor from the mountain crests.

Napoleon marched up the valley of the Durance on his way to Grenoble. The idea of invading France through the Durance was to avoid the Rhône area, which was hostile. After his escape from Elba, he came by way of the Alps, first landing at Golfe-Juan, then cutting through the limestone mountains.

Military strategy does not intrigue McBride. Other than badminton — and a little tennis to accommodate me — he scorns sports.

I tripped on a limestone boulder. I lay on my stomach, not hurt or exhausted, but in an attitude of contemplation. At eye level were assorted bivalves in the limestone, the valves unbroken and immensely varied. I would have to get up sometime, but I took my time brushing sand from the boulder and my face, gradually pushing up and rising above the bivalves.

McBride nimbly crossed on the rough ground ahead, filled a sample bag, and wrote a number on the tag. I saw something sticking out of the sand between two boulders. It turned out to be a brown slipper, cracked at the toe and curved back like a boomerang. Abandoned shoes are abundant along stream courses and highways everywhere.

Half an hour later we drove up into the mountains and the switchbacks. It began to rain. We passed a man carrying two long rods and a paperback book. Two black dogs walked ahead of him, tracking a hundred head of white sheep, a black one — perhaps a counter from a larger band — and two orange goats.

The mountains bore marks of their history — cirques and rock steps, arêtes and horns, the scarred Alpine bedrock. We passed a tarn and, below, could see a hanging valley partly filled from the rain and cascading into a landslide lake. The man and the dogs stopped at the edge of a small meadow under the trees. In the rain the isolated rocks were dark islands.

Why glacial theory was so slow to evolve became a mystery to me after I spent some time in Switzerland. Glaciers still fill high valleys. Polished outcrops bear striations everywhere. The ice-cut surfaces are among the most spectacular in the world. According to Anthony Hallam, many uneducated Swiss peasants believed in a version of the glacial theory early in the 1800s. Why not? Our shepherd had probably correctly interpreted much of the evidence he walked over on the mountain and in the valley.

In the fierce rain, the windshield wipers faltered. I rolled down the window again and saw demoiselles moving down the mountain front, their hats inclined erratically and shedding sheets of water that fell at many angles. The tall columns of glacial till, capped by the large boulders, had chosen a poor time for an evening stroll.

We drove on toward Serre-Ponçon. The rain stopped abruptly. A pair of rainbows hung over the Durance between the mountain walls. Vertical chutes of water rained from the hanging valleys.

The small town at Serre-Ponçon offers one place to stay, the Hôtel de la Poste. The reservoir at Serre-Ponçon has a capacity of 264, 200 million gallons, but the hotel does not routinely bring water to your table in the dining room. Red wine is abundant and we had a local wine with vegetable soup, veal, rice, stewed tomatoes and lettuce, the only dinner served at the Hôtel de la Poste.

The only other people at dinner were a young Swiss couple. They drank a rosé which the man heartily endorsed, drinking most of the bottle and writing down the name of the winery. The young woman picked at her food but joined her companion for dessert, a large gaudily wrapped rectangle of banded chocolate, vanilla, and strawberry ice cream. The ice cream was too large for the saucer and dripped on the tablecloth. The man eagerly set the woman straight on the wine, the dam, life, and dripping ice cream.

We did not see them at breakfast. Several regulars were having coffee and brioches, and it was not too early for one of them to have a glass of red wine at the bar. I ordered hot chocolate, which was delicious with the fresh bread and apricot jam.

Near the hotel is a fine bakery, said the woman serving breakfast. After wandering along several wrong streets, we found it where she said it would be. For lunch we bought a long loaf of bread and, from a little store nearby, ham, tomatoes, cookies, and bottled water.

White clouds had settled below the crest of the mountains — early morning clouds that veiled the flanks and hid the mouths of valleys. It had rained in the night. The grass beside the road was bent and arched toward the ground-green roller coasters with drops of rain poised to rush off the crest. I loaded the car. McBride paid the bill.

The traffic on the bridge over the Durance at Noves picked us up on the north side and carried us turbulently to the south side far beyond the river. We did not intend to go so far, preferring a laminar pace near the ground, moving close and straight without turbulence. We were held on the road in the shooting flow of traffic, unable to deposit ourselves where we wished. McBride's great driving skill and the stability of the Renault saved us from being tossed up and back in the traffic's torrent, an antidunal artifact inclined upstream like casual gravel perversely imbricated.

Far past the river we flew off into a little woods. We crossed the road and joined the traffic heading north. I looked for sand bars along the river; McBride looked for a safe turnout. So we went back and forth over the Durance.

I said I saw sand along the north bank. McBride doubted it, but would not bet on it, and turned onto a side road. There was no sand. I had mistaken foam and froth on the water for a marginal sand bar.

The woman at our hotel in Arles said we were too late for dinner. She was by herself at the bar and at the desk. Though it was only eight o'clock, the chef was gone, she said. She had customers in the bar. She recommended a restaurant down a narrow lane. We sat 20 minutes without service and decided against eating there.

We ate in the center of Arles on a cafe terrace, the daytime smell of cars and trucks still in the air. The outside of the cafe was brightly lit, the light harsh in the starry night and severe on the white plastic table. The legs were wobbly, sometimes trembling and moving capriciously. There was little traffic through Arles, and no one else eating on the terrace.

I remembered a night at a little bar in Manderson on the Big Horn River in northwest Wyoming. The bartender had turned off the outside lights and drawn the blinds. It was two o'clock. The orange and green glow of neon lights over the bar and the blood red light from the juke box clashed in the dreary room. The ordinary faces of the customers turned into those of hooligans. They matched for silver dollars, drank beer, and told war stories, which by then — a year after the end of World War II — they were retelling to the same people.

I looked at a picture on the Arles cafe wall. It was spotted and stained, a print of Cezanne's *La Montagne Sainte-Victoire*. It did not belong where strangers dared not stay for long.

I ate the runny omelette quickly and drank a glass of white wine. McBride talked about the Rhône River in Switzerland, the undercurrent into Lake Geneva, and the subaqueous forms of

the Rhône delta there. The river arises partly from glacial melt-
waters. It is colder and contains more sediment than the lake
water. The differences in density produce turbidity currents that
flow down the delta front to depths of more than 900 feet. We
decided to go to Port-St-Louis, one of the places where the Rhône
enters the sea in France.

Late on Wednesday evening we drove south from Arles, following the Grand Rhône, to sample the beach at Port-St-Louis.
Could we recognize sand that had come from the Durance at
Briançon, grains formerly bouncing and rolling from Serre-Ponçon
to Sisteron in some remote flood before the river was tamed;
carbonate grains off the Lubéron entering the Durance at Cadenet;
quartz running to Noves and Avignon, south to Arles in the Rhône,
and down the distributary channels to the beach? Unlikely. Yet we
drove across the Camargue toward the port.

Each year the Grand Rhône, which contains 90% of the flow,
carries 20 million cubic yards of gravel, sand, silt, and clay toward
the Mediterranean. Paris would be covered by 10 inches of alluvium every year if the Grand Rhône debouched on it, said
McBride. The image pleased him.

The delta is extended 30-50 feet per year. The lands of the
Camargue have been retrieved from the sea by the building of
the delta.

On our way to the beach through Port-St-Louis, we wandered
along many wrong roads. The mouth of the Grand Rhône is now
defended by poor maps and a few confusing signs.

McBride, driving by instinct, found a road to the sea. In the
calm evening the beach was deserted except for a man setting up
a metal rack of cheap curios beside a group of a hundred or so
folding chairs. The bareness of the beach, the lack of any debris
on the ripple marks, and the man's leisurely pace seemed to say
that the chairs were for a meeting yet to come. The sea was
green, but in the evening light, had taken on a tinge of pink and
gray.

Across the bay we could see Marseilles. The long stoss sides
and the short lee sides of eolian ripple marks on the dunes gave
the general wind direction — onshore toward Arles. Water had

broken through the crests of marine ripple marks, cutting paths on its way back to the sea. In a seaward channel, we found a row of yellow onions. Farther on, onions covered the sand.

McBride and I sat mulling over rivers we had known, the low pay of teachers, the harm television has done, and the high cost of practically everything in France. I was sipping a dollar-and-a-half cup of coffee and taking tiny bites of a two-dollar brioche while he studied the maps, tracing our likely route from Arles down the coast into Italy. We wanted to be in Portovenere in a day to meet R.L. Folk, compulsive Italian traveler, first-class talker, professor, amateur painter, and one of the two or three best sedimentary petrographers in the world.

I looked at half a dozen post cards I had bought for about 15 fr. at a book shop — Van Gogh's *The Yellow House*, *The Blue Cart*, *Vines*, and several others. Van Gogh had come to Arles from Paris on February 20, 1888, and had rented the yellow house the last of April for 15fr. a month, about $2.30 at the current rate of exchange. *Sunflowers*, one of his many paintings in Arles, sold for $40 million in 1987.

Van Gogh was not a painter of fountains. I looked in the book shop among the works of the impressionist painters for a fountain or a waterfall, even for a stream cascading over boulders in a mountain reach, but found none. A good friend shares my affection for fountains — I had hoped to send her one from Arles. No fountains.

I scribbled a few lines from a Baudelaire poem ("The fountain leaps and flowers/In many roses,/Whereon the moonlight flares.") on the back of a card. I wrote about the Penitents of Les Mees, sandstone-cemented conglomerate columns more than 300 feet high beside the Durance, below Sisteron. During the Saracen invasion, the monks of Monte Lure were enraptured by some Moorish women a local nobleman had brought back from a campaign against the infidels. To punish them for their longing, St. Donat petrified the monks as they walked single file along the river in their pointed, hooded cloaks. We ate lunch in the shadow of an especially tall monk.

For a mere $16.00 each a night, our hotel in Arles, a shabby old inn on one of the principal streets through the city, rented us a smaller and less sumptuously appointed bedroom than Van Gogh has immortalized (*Bedroom at Arles*). No pictures graced our walls, or at least none that I could see in the light of the single bare bulb. Shaving was hazardous. Our shuttered window faced the street and, fortunately for us, I now believe, it was locked fast. The exhaust fumes that did get in from the early morning traffic were less intense than they would have been if we could have thrown the window open for air.

After we had spent half an hour in the cafe, McBride's eyes began to blur. He could have no coffee — a sensitive stomach — and my delight in post cards mystified him. When I picked up *Le Pont de Langlois* again, he folded the maps and took off down one of the tortuous streets to buy some pears and the only cookies his stomach now permits — vanilla without creme fillings or glaze or powdered sugar on top.

I put *The Blue Cart* on top of the other cards. Art historians have given other names to this painting (*The Harvest,* and so on), but these I do not like. I looked at the card again, and recited Williams' poem to the empty cafe:

> *So much depends*
> *upon*
>
> *a red wheel*
> *barrow*
>
> *glazed with rain*
> *water*
>
> *beside the white*
> *chickens*

We collected a bag of sand at Saintes-Maries-de-la-Mer. Perhaps, I thought, we were near the spot Van Gogh painted the little wooden fishing boats: "On the perfectly flat, sandy beach little green, red, blue boats, so pretty in shape and colour that they made one think of flowers."

But perhaps not. There were many pot-bellied red, white, and gray tourists and pilgrims on the beach. It was hard to sustain the belief that Van Gogh had been there.

We crossed the Camargue to Saintes-Maries-de-la-Mer on 9 June 1988, 100 years after Van Gogh painted the fishing boats and the old church. The sea was flat running. Strand bars stretched into the distance — there were no folding chairs or onions on the sand.

I stared out to sea. The sky was flint blue with a great bright sun. McBride disappeared down the beach line, looking for a pristine place to collect sand.

The saintly ship came into view, two Marys standing in the bow. The Provencal legend: Mary, the mother of James "the less"; Mary Salome, the mother of James Major and John; Mary Magdalene and Martha with their brother Lazarus; St. Maximinus, Suedonius and Sarah (the black servant of the two Marys) were abandoned in c40 to the waves without sail or oar by the Jews of Jerusalem. Miraculously, their boat carried them across the Mediterranean to bump up gently on the sand at Saintes-Maries-de-la-Mer. Their joy was great. They built a simple oratory on the beach to the Virgin. The disciples then separated — Martha to Tarascon, Lazarus to Marseilles, Mary Magdalene to the Ste-Baume, Maximinus and Suedonius to Aix. The two Marys and Sarah stayed in the Camargue and were buried in the oratory. Pilgrimages to the saints' tomb commenced at once.

I plotted the site on the map. We drove toward Tarascon to collect the Rhône sand, our last stop. Lean black bulls grazed in the pastures. In the distance, and tied to hitching posts near the road, white horses browsed and stamped the ground.

With the sign of the cross, Martha vanquished the Tarasque, a monster that climbed out of the Rhône to devour children and cattle and prey upon anyone who dared to cross the river. In the spring runoff, the Rhône was high and brown, covering the sand bars.

We stopped south of Tarascon and walked toward the river. There was a shoe in the grass. We could not see sand anywhere — there was no chance to sample in the spring. Should we have asked for Martha's intervention? McBride suggested instead that we ask Medard Thiry, a French geologist working in Fontainebleau, to help us in the fall.

On the Beach at Elba

Happy are those who can tie together in their thoughts the past, the present and the future!

— Alexis de Tocqueville, 1858

In June of 1987, disregarding a one-star rating in the Michelin guide to Italy and a pinched bank account, I took a car ferry from Piombino to Portoferraio, went to 20 different beaches around the island of Elba, and came away with 20 sacks of sand. To a geologist on the move in the spring, Elba is not a one-star tourist attraction but rather the mysterious remains of a spectacular mountain range, part of Tyrrhenia, a land that vanished beneath the sea.

The impulse to move and to keep moving is strong after a long winter in academe. The snow has melted; the rocks are uncovered. I want to see strata I have not seen before. On Elba, I want to collect fragments of former worlds that have been carried by slope wash and streams to the beaches of the island.

"Never travel alone," said Montaigne, or was it Samuel Johnson? I nearly always follow that advice. On this journey, like so many others, I was fortunate to be with Earle McBride, accomplished student of sedimentary rocks and speaker of Italian. His wanderlust equals mine — perhaps exceeds it. He is comfortable with the conditions one finds in the inns of small Italian towns: beds of extreme concavity, unsure showers, no soap, no wash cloth, a tiny towel, faint lights, and the weathered Gothic church whose rusted bells chime each hour through the night.

We went to Ilva — named after the original Ligurian people — to study the history of the beach sand. Has potassium feldspar — a less stable mineral than quartz — been more highly rounded? Has quartz — of common minerals the most resistant to chemical and mechanical weathering — been weathered at all? Do the minerals of the beaches, in their occurrence and abundance, reflect the composition of the rocks and sediments from which they eroded? These are questions that a certain species of geologist — the imaginative sedimentary petrologist — seeks to answer and no one else entertains.

We thought the Tyrrhenian Sea had been approximately at the same level for 5000 years. We were driven to look at grains smaller than muffin crumbs to piece together a history whose artifacts had slowly diminished as they were shifted up and down the beaches and along the shorelines.

The night before we left for Elba we stayed at Massa Marittima, once near the sea but now 10 or 11 miles inland. In beginning geology courses, teachers say the sea came in and the sea went out. In Italy, the distance from mountains to seashore is small. The sight of a troubled and noisy sea coming in would not have surprised me.

The Fauls, a husband and wife team of historians, noted in 1983 that Giovanni Boccaccio (1313-1375), from shells found inland in Tuscany, had recognized the remains of former sea life. The marine origin of such fossils was certainly also clear to da Vinci (1452-1519), said the Fauls, and da Vinci "... wrote it down in his notebook and went on to other things." He was the first to think it through and get it right.

On our way to Massa Marittima, we paid four dollars per gallon for gasoline, a price we may soon pay in the U.S. if we continue without an energy policy. We did not get even four dollars' worth. Massa Marittima had closed early. The staff went home for lunch and stayed, said McBride.

We drove in and out and circled the town. No hotels. When we finally found one it was far down the hill slope — near, it seemed, to one of the iron or copper or pyrite mines.

At dinner a fly took a sip or two of McBride's red wine and skittered toward his tortellini en brodo. The cafe was dark. A fly could easily drown in the soup unnoticed, said McBride. He suggested we walk to the Piazza Garibaldi to see the cathedral, one of three Romanesque buildings in the square that were built in the thirteenth century.

I was very tired. It was a relief to sit on the steps and look at the roaring lions on the facade, the blind arcades, and, in the dim light of the square, the paired windows of the Palazzo Pretoria and the crowns of merlons.

We bought ice cream cones and joined the young people on the steps. I ate ice cream and looked at the lions, their tired old faces transfigured by weather and pigeon droppings. The young people looked at each other.

I thought of Philadelphia, the Liberty Bell, the great flocks of pigeons there, and of dodging the swooping feeding birds. In the first few months of a short tour in the Navy, right after World War II, I was stationed near Philadelphia at Swarthmore College. Philadelphia was the largest city I had ever seen.

Where are the descendants of those swooping pigeons? What has become of the young man who lugged *Geology, Volume I, Processes*, the classic text by Chamberlin and Salisbury, out to the Naval yard? I read about "The Origin and Descent of Rocks" at a time when I wondered if I was rising or falling. As I looked over the square at the dark Romanesque building I asked myself rhetorically if I could tell yet. I finished my ice cream. We drove back to the hotel and prepared to leave early the next morning to drive to Piombino.

A boat for Elba. There are only two a week, or sometimes three. The schedule seldom matches the actual leavetaking.

We drove to Piombino late on Thursday afternoon anyway to buy a boat ticket for Friday morning The four young girls behind the counter, fluttering like finches feeding, flew from computer to computer, amused by, and amusing to, the Americans. Finally, they sought help from a young man, the blue jay in that little office flock.

He spoke — they sold us a ticket. The girls laughed and waved and settled behind the green computer screens as we prepared to leave. The blue jay strutted down a long aisle behind the counter.

During the night McBride had an attack of anxiety. "We have a ticket to buy a ticket," he said early on Friday morning. When we got to the dock, he went to the office of TOREMAR. No boat ticket. He stood in line for another hour while I sat in the car, ready to move it out of a restricted zone.

We parked the car below decks. The diesel smoke of the mighty trucks enveloped us. We went up on the deck and stayed there for the whole trip, away from a howling horde of children, the German and Japanese tourists on their way to sand and sun, and the moldy smell of the close cabins.

It was a sunny morning. We put out from Piombino, and the iron works belched suffocating clouds of acrid smoke and dust over the town and the harbor. At sea, I forgot about the dust settling on the city and farm land.

As we approached the Elban coast, the children and the Germans and Japanese crowded onto the deck. They arrived for group photos and a shot of the steep rocky cliffs. In places the sea edge was discouragingly barren of sand. An old Italian woman shouted at one of the children — words strong and short. McBride was not certain whether she was swearing. He has not mastered that part of the language. The boat docked with a delayed bump, and we hurried down the stairs to the car.

We lunched at Marciana Marina by the water's edge. Before we decided on a cafe we looked at the menus outside the restaurants and walked back and forth along the promenade, enjoying the sea, the oleanders and magnolias, and the palm trees.

Our cafe was one of the least expensive, about 25 dollars for the two of us. We ate little grilled fish, as we did on Elba wherever we ate. We noticed the great height of a young woman sitting near us. McBride said she could be a star in badminton, but the waiter said she was a star in volleyball, a former member of the Italian national team, an attainment Italians claim for many. I could find no word for badminton in Italian. I doubted that badminton had become a power game in Italy as McBride claims it has in Austin, Texas, where he plays.

We drank the Elba bianco, a wine McBride made few notes about, but one I found pleasingly dry. We asked about the famous perfumed wines, the white Moscato or the red Aleatico, but they had none. Giani Zuffa, our friend in Fontanelice, later said these wines are hard to find even in Bologna, where he teaches at the university, and they are expensive.

While we ate, a large contingent of Italian naval officers passed by and, after much discussion, took a table reserved for them in the restaurant east of us, the most expensive eatery along the promenade. The United States government, through the contributions of its citizens, pays for these lunches by funding the North Atlantic Treaty Organization, says McBride.

After lunch we sampled the beach sand. The streams that drain the highest part of the island — the 3340-foot granite massif of Monte Capanne, composed of 7-million-year-old granite and granodiorite — flow into the bay and deposit sand in a delta. Our sample, later studied with a petrographic microscope, reflects the composition of a granitic source terrain in its grains of rocks and minerals: quartz, 51%; potassium feldspar, 19%; plagioclase feldspar, 5%; igneous rock fragments, 10%; metamorphic rock fragments, 7%; unknown rock fragments, 3%; carbonate rock fragments (from the beach), 2%; miscellaneous (a category reserved for grains we have not identified), 3%.

A pinch of sand from a beach may be enough for a geologist to describe the rocks of a distant highland, rocks he has not mapped or visited.

On the gulf of Marciana Marina is a small jetty with a circular stone tower built in the eleventh century by the Pisans as a defense against the Saracens. It became the property of the Medici, thus confusing guidebook writers. Is it the Pisan tower or the Medici tower? In mid-afternoon, on our way to the west, we stopped to look. Below, dark figures danced in the green sea, and the narrow beach was a quarter moon tied to the headlands.

Italians are prolific road builders, and there are many roads on Elba. Napoleon is given partial credit for this. In his brief stay

there, he restored and embellished his town and country resi-
dences, reorganized the army and navy, built a theater, con-
structed some of the island's most important roads, and invented
a new flag, a red strip with three industrious golden bees on a
white background.

From Marciana Marina to the west, and around the high western end of Elba, the roads wind and are intricately switchbacked
— intestine-like trails on the massif and the metamorphic ring of
serpentine rock. The number of vehicles rocketing out of the
curves on the wrong side of the roads is large.

We took a sample from a pocket beach near Zanca and another
from the sand at the foot of Punta Nera, the westernmost point
of land on the island — a Jimmy Durante nose silhouetted by
the sea. The beach at Punta Nera is three-fourths serpentine and
pyroxene–serpentine grains of pale green, yellow, and gray.

No one else had come down the clean-sliced cliff at Punta
Nera in a long time. Or so it seemed. The quiet of the shadowed
bay, the strong winds blowing against the Durante nose, the narrow band of green sand, the warmth of the mid-June sun at the
sea edge, the warble of hidden finches — a feeling of coming on
an ancient shore. I was suddenly happy.

Along the bold cliffs on the southwest side we took turns going
down to the beaches for samples. On McBride's turns, nudists
invariably appeared near the hummocky dunes as he poked along
the shore. He didn't see any of them, he said, his sights being
on the cross strata of the dunes, the planar bedding, and mica
schist of the splintered and sheared metamorphic ring. "How do
you see crystals of biotite, a sixteenth of a millimeter in diameter,
and miss a naked blonde?" I asked.

On one of my forays I cut my finger on a sharp schist parting.
I was surprised by how much it bled in a few minutes. I came
to a cliff face too steep to traverse and retreated, looking for a
gentler route. By then the blood had clotted.

When I was a boy I was afraid of everything — of falling, of
drowning, of being bitten by a rattlesnake, of large children, of
the dark, of shadow figures cast by sandstone. My fear of heights
still comes and goes. I do not look down from balconies or from
the tops of tall buildings. I dream of terrifying heights and of

falling. But most of the time I somehow reach the peaks I need to attain in the field.

In a bar at Capoliveri, two men dressed in rough working clothes drank cappuccino and argued about the Italian National Soccer Team, recalling days of glory, the World Cup. They had many suggestions for the coaches. They did not believe that the team, in its attack, had the ability to score on England, a team whose defenses they did not respect.

The woman running the bar, a capable and outspoken woman in her middle forties, accused the players of arrogance, of being spoiled by wealth, and of laziness. "The team manager is out of touch with modern soccer," she said.

I searched the newspapers for results of the NBA playoffs. Had the Boston Celtics pulled even with the Los Angeles Lakers? Had Larry Bird recovered from his sore feet? McBride, who believes the games are fixed, is little help with basketball scores in the Italian papers.

I was slow-witted that morning: it took me 10 minutes to find the article and read it. Los Angeles had won again. Magic Johnson had played well. I told McBride. He said the games were fixed in Las Vegas. It is all just business.

The two men left. Some children came in and began playing video games. McBride drank another glass of milk and ate a roll filled with yellow pudding. Outside, we drove along to where we could stop and see the Three Seas — the bays of Portoferraio, Marina di Campo, and Porto Azzurro. It had rained in the night, and the air smelled of wet clay. There were no clouds. The lapis-blue sky sharply overlay the turquoise-green seas, calm in the distance and flat-running. There was a starving, pregnant, orange cat hunting beside the road.

South of Cape Pero, near the mouth of the Rialbano on the eastern coast, we collected grains from a black-sand beach.

The sorted and rounded blood-colored grains of hematite (red ocher) and the black magnetite (loadstone) and ilmenite (titanium ore) are eroded from highlands on the west, where these minerals have been mined since the Iron Age. The river carries them to the beach and to a delta. It is a good place to see the shoreline, a relatively straight segment from Rio Marina to Capo Pero. We zig-zagged our way along the sand like coarse clasts caught in oblique waves moving erratically along the beach. The dunes and shore were littered with the garbage of sun worshippers. There were only a few people on the sand. The Verrucano Group, metamorphosed Triassic strata we had studied at Monte Argentario, crops out to the west and along the coast to the north. The origin of the sandstone is as enigmatic on Elba as it is at Monte Argentario. There was no breeze, but I smelled the sweet scent of broom.

I walked along the berm. A flock of sparrows flew up, swooped toward a low rock promontory, then slid sideways in the air back into the trees bordering the stream. I wondered if my son was remembering to fill the bird feeder. There was a sense of endless space in the long line of the berm.

I dug into the black sand. In the high bright light, the specular hematite shone metallic and splendent. I let a few crystals fall through my fingers. They caught the sun on rhombohedral faces, flashing fire as they fell on the dune face and rolled seaward. My fingers turned a lurid red, the hands of Diggory Venn.

I rubbed a few grains across a piece of porcelain. They left a streak — reddish brown, cherry red, and brick red. At the edge of a slender crystal, brilliant blood-red light flashed down and across my lifeline. In the tiny grains of the beach sand, grains a sixteenth to half a millimeter in diameter, there were crystals as beautiful as any in the world's museums.

The specific gravity (the weight of an object divided by the weight of an equal volume of water) of iron oxides is high (4.5 to 5.2) compared to that of quartz (2.65), the dominant mineral in most beaches. We collected a sample at the mouth of the Rialbano, which contained 97% opaque grains — practically all of them hematite, magnetite, and ilmenite. Their concentration on the beach in such overwhelming amounts is the result of their abundance in the iron-rich highlands and their high specific gravities, which lead to selective sorting — light minerals moved out, heavy minerals left behind.

McBride wished for a platinum placer, such as occur in small amounts as beach sand in southern Oregon, and in large amounts on the east slope of the Ural Mountains. A geologist can always imagine a more exciting deposit somewhere else.

We collected our last bag of sand at Bagnaia. From the little horseshoe-shaped bay we could see across the large bay of the harbor to Portoferraio. We sat on the sand drinking bottled water, eating ham and tomato sandwiches, and looking out to sea and into the harbor. Cretaceous strata have been folded into an anticline at Punta Pina, and the gray limestone limbs dip steeply. At noon there were no nudists. We sat silently watching the waves break parallel to the shore. As I watched, I fancied I saw Napoleon, no longer a king, disembarking at Portoferraio, capital of the meager territory which was his only possession. There was despair. There was pride. His practical eyes took in the terrain, preparing for a skirmish, planning his escape. Gulls dived under the prow of his little brig.

We finished our lunch and drove slowly toward Portoferraio. The bag of sand was wet. McBride put it outside on the hood, tied to a windshield wiper. The boat was at the dock.

Later, when I looked at the sand in Salt Lake City, I found it contained grains typical of granitic terrains — monocrystalline and polycrystalline quartz, potassium and plagioclase feldspar, igneous and metamorphic rock fragments — sand very much like the sand at Mariciana Marina, though unlike the sediment and rocks near Bagnaia, or in its drainage area. Had the beach at Marciana Mariana been carried to Bagnaia while we slept?

Through the Dolomites and Apennines

The Apennines are a curious part of the world. At the edge of the vast plain of the Po, a mountain range rises from the lowlands and extends to the southern tip of Italy with a sea on each side. If this range had not been so precipitous, lofty and intricate that tidal action in remote epochs could have little influence, it would be only slightly higher than the rest of the country, one of its most beautiful regions, and the climate would be superb. Instead, it is a strange network of crisscrossing mountain ridges; often one cannot even find out where the streams are running to.

— Johann Wolfgang von Goethe, 1786

Having spent the sweetest days of spring in a stuffy interior room with no windows teaching reluctant students about sedimentary rocks, I left on a Sunday in June to treat my numbed mind and infected sinuses to mountain air in the Italian Dolomites and Apennines. I first knew of the Abruzzi — the most mountainous part of the Apennine range — from reading Ignazio Silone's *Bread and Wine* when I was 19. At the time I thought I would never get there, my days then being devoted to manning an officers' mess at the Great Lakes Naval Training Station. After reading Silone a geologist wants to see "the end of the world," a village destroyed twice by floods and once by earthquakes. Or to walk in valleys "split, cracked and bereft of vegetation." To a geologist, Silone's Abruzzi resembles the carbonate terrains of western Ireland, the high glaciated mountain valleys in the Big Horn Dolomite of Wyoming, and the stark, rock-striped, shattered Paleozoic ranges of eastern Nevada.

As for the Dolomites, I had fallen for them the year before when Earle McBride and I — taking a great jog to the east — passed through on our way to France. Golden dolomite, crowned and ribboned with snow streams, is a rock I would like anywhere.

At the airport in Salt Lake City I lost a quarter making a phone call. Next to me a young broadjumper from Louisiana State University called to tell her mama that she did 19'7" on her first jump at the NCAA Track and Field Championships. "I knew I could," she said. "We won. U.C.L.A. was second." After a long pause: "It's in California."

I took a plane from Salt Lake City to Minneapolis and then to New York City. A careening carry-on cart in the overhead storage pierced the mouthwash I carried in a small bag. I spent 2 hours in New York trying to clean the bag, whose breath was sweet but whose bottom was greasy. We flew on to Zurich, arriving at 9:15 Monday morning. Earle McBride, sedimentary petrologist from Austin, Texas, was waiting. He said that I smelled strongly of peppermint.

From Zurich, we headed for Altdorf on our way to St. Gotthard Pass on the north side of the Alps. We were too early in the year — the pass was closed.

A statue of William Tell stands in the Altdorf main square. A postage stamp bears the likeness. Our mental image of William Tell comes from the statue and stamp.

We lunched in Altdorf before heading east for Innsbruck. We bought pastries at a bakery. In Innsbruck that evening after dinner we walked for an hour and a half up the mountain above our hotel and into low clouds near the crest. It was not enough to calm all of the French fries, said McBride.

Most of the clouds had cleared by 3:30 in the morning, leaving only a few thick white ones below the high terraces under the forest line down near the river. I could not sleep — it was 8:30 in the evening in Salt Lake. By 7:00 a.m. the clouds were back and the morning was overcast and as cool as it had been the day before.

In a few hours we crossed the border through Brenner Pass and were in Italy. We began to sample the sand bars along the Rienza. A road-crossing guard in a little town asked us what we were doing near the river. McBride said we were sand collectors.

We drove back and forth across the Dolomites and across the Apennines at many places. We walked along the major stream courses and sampled rocks of Apennine valley walls and peaks. We drove from Cortina d'Ampezzo in the Dolomites to Ravenna and south along the Adriatic Sea to Roséto and then to L'Aquila and through the Abruzzi. We stayed a night in Pompeii. We went to Tivoli east of Rome. We went back into the Abruzzi. We drove northwest and north to Viterbo, Gubbio, Bagno di Romagna, Modena, Fontanelice, Loiano, and finally northwest from Bologna to Salsomaggiore Termi on the A1. And then we crossed into Switzerland through St. Gotthard pass, recently opened. The whole trip took 17 days. We stayed in 14 towns and cities. On such a short trip there was much we did not see.

The Dolomites were named to honor the French geologist, Déodat de Gratet de Dolomieu, Knight of Malta, captious adventurer, aristocratic adherent of the Revolution, august geologist. Dolomieu studied the Dolomites at the end of the 18th century. He studied Italian volcanoes, and knew that the heat could not come from combustion, but did not reach an alternative explanation. The common calcium–magnesium–carbonate mineral dolomite, whose reaction to dilute hydrochloric acid is to form slow-breaking bubbles, also honors Dolomieu as does the rock dolomite, a stone with 50% or more of the mineral dolomite. It is imprecise to give mineral, rock, and mountain range the same name. But geologists, who are little honored anywhere and sometimes ignored among more mathematical and experimental scientists, can rejoice in honoring the dead Dolomieu.

The Rienza River — roughly the northern boundary of the Dolomites — was running high. We could not collect samples of the point bars as frequently as we wished, but we bagged dolomite sand where we could along the river. We ate lunch on an old toll road beside the river, then drove up the Rienza and admired the valley profiles. We stopped at the summer villa where Gustav Mahler composed his Ninth Symphony. On the sunny valley slope, with clean, unglaciated dolomite peaks forming the skyline and splashes of spring wildflowers coloring the floodplain, the day did not evoke the shadows of Mahler's Ninth and his early death at 50.

Crotina d'Ampezzo is the capital of the Dolomites and the largest town with a population of about 10,000 people. The Olympic winter games took place here in 1956. It is a resort town still. The town sits in a valley beside the Boite River, the major tributary of the Piave, made famous by *A Farewell to Arms*, at an altitude of 3950 feet. On the west stand the three domes of Tofane, the highest at 10,640 feet. An ossuary near Cortina holds the bones of 10,000 soldiers killed in World War I.

McBride bought the first of many maps he would buy in Italy. He cannot pass up a slightly different map. "This map will show the locations of small towns we have stopped at while studying rivers and rocks," said McBride.

We walked around Cortina in the chilly evening air waiting for the restaurants to open at 7:00. There were many stray cats and few people, Cortina being between seasons. The cats looked healthy, unlike the animals on the dock at Portovenere.

In the morning we took a sand sample from the channel of the Rienza headwaters. We were alone in the ancient mountains. We saw the Dolomites as Neanderthal man may have seen them — snow plumes blowing from the high peaks stretching out toward Africa and, in the upper turbulence, the plumes reversing and sometimes swinging northwest toward France. The feeling blotted out the evening's queasiness.

After collecting the Rienza sand, we looked at our map of the Dolomites, counted sample stations, and decided to sample another river draining the range. The Rienza seemed unpromising. We went to the headwaters of the Boite, where it begins its run south, and collected a bag of sand from this shallow clear tributary to the Piave. It began to snow. Dolomite mud near the banks became pitted — miniature rimmed impact craters where rain and sleet and snow splashed down.

McBride had brought an umbrella. It was snowing hard when he opened it.

It was gray and overcast at noon when we reached Longarone. The cemetery lawns had been recently tended. There were many cut flowers on the markers — red and white carnations and a few vases of spring wildflowers. Rain had fallen off and on all night and began again, lightly. Drops of water started to collect and run down photographs of people killed in the catastrophic flood that swept over the crest of the Vaiont Dam east of Longarone on October 9 of 1963. I had never seen a graveyard with so many pictures of victims, or so many of them children. I walked among the markers, sometimes speaking a family's name into the light rain.

At night on October 9, about 90 million cubic feet of mountain slid down steep slopes at 60 miles an hour into Lake Vaiont, a deep reservoir behind a dam in the Italian Alps. The reservoir was less than half-filled. When the slide dropped into the reservoir the water rose 787 feet above its previous level. The slide material filled the reservoir valley for more than a mile along its axis, and to heights of 500 feet above the reservoir level. The slide and accompanying blasts of air and water and rock caused strong earthquakes recorded many miles away. It blew off the roof of a man's house more than 800 feet above the reservoir and showered the man with rocks and debris. A great wave broke over the dam, 300 feet above its crest, and fell into the gorge below. More than a mile downstream, the waves were still more than 230 feet high. Constricted by the steep-sided valley, the water increased its velocity tremendously, snatched up mud and rocks, plunged down the valley, and burst across the wide bed of the Piave River and up the mountain slope. In Longarone and adjoining towns, water, mud, and rocks killed about 2600 people. It was over in less than 7 minutes. It was the world's worst dam disaster, though the dam stood and the main shell and abutments were not damaged.

Extensive efforts had been made to stabilize the jittery slopes above the dam. For the three previous years, movement on the slopes had been monitored. Engineers assured everyone that the dam was safe and that the right things were being done. Animals grazing on the slopes doubted the engineers and moved away a week before the flood.

One shoulder of the dam was supported by Monte Toc, called *la montagna che cammina* by the local people — the mountain that walks. There had been an ancient landslide near Casso above the dam site. In 1960, opposite the Casso slide on the south side of the valley, there had been a smaller slide into the reservoir. Engineers expected another, smaller landslide. Until the day before the disaster, they did not realize that a large piece of land was moving as a uniform mass downslope.

Geologically, almost everything was wrong with the site. The instability of the land was evident. The valley was steep-sided, sharply undercut by the river, with dizzying slopes. The rocks — limestone and claystone and slippery clayey interbeds of the Jurassic Lias, Dogger, and Malm and of the Cretaceous — were inclined toward the valley axis, providing about the worst stratigraphic and structural geometry for a dam foundation and reservoir. The fractured limestone was riddled with underground solution caverns that collected water and added to the ground saturation. Surely geologists would have noted these relationships and spoken about them before the dam was built. From the outside, one might conclude that in Italy, as everywhere else, dam engineers don't pay much attention to geologists.

Before the landslide, downslope movement of regolith — called "creep" — had been 0.4 inch per week, a rapid rate. In September it increased to 10 inches per day. A day before the landslide the rate was about 40 inches per day. Beginning in late September the rains began to increase the weight of the unstable slope, raising water pressure in the rocks and producing runoff that raised the reservoir level while the engineers were trying to lower it. Water from the reservoir seeped into the unstable layers. The great block broke away.

Later in Salt Lake, Chad Gourley told me the Italians sent the engineers to prison after the Vaiont flood. He did not know whether or not any geologists accompanied them. Geologists and engineers studying dam sites seldom consider the prospect of prison in the course of their studies.

We filled the car with gasoline — about $3.50 per gallon in Longarone. I had a cappuccino at a small bar, where a dusty dispirited dog sat on a small mat near the door. McBride was still off coffee and his stomach was in revolt over some spice or other from dinner. We drove south, staying close to the Piave and looking for a place where we could get out on a sand bar to sample the sand. The river was flooding, and many sand bars were covered. The light rain suddenly became a squall, and then it stopped as suddenly. The sun came out. The river was brown and heavy with sediment, dirty and heavy and bearing no resemblance to the clear blue-green waters running over the bent and broken dolomite in the headwaters above Cortina.

That night we stopped at Ponte della Priula, a tiny town near the Piave, which was absent from many of our maps. McBride's stomach had regained its stability and we had pasta with salsa di pomodoro, caponne, and a tossed green salad for dinner. We drove north 10 miles for dessert. It was a neighborhood hangout for young men and women. They came in pairs on motorbikes and on foot for a pastry, a cappuccino, and a campari. On the way back, McBride, now tired after several days of the piano music of Erik Satie and Saint Saëns's Third Symphony, pushed in a new tape: music of Lucio Battisti played by Richard Clayderman. Rushing between rows of cypresses with the car windows down and the sentimental music washing over me brought memories of summer nights in high school, the 30-mile drive from Worland, Wyoming, to Thermopolis, listening to Peggy Lee, Nat King Cole, and Louis Armstrong, troubadours of a love that seemed just beyond my reach.

Late in the morning we arrived at Punta Sabbioni, south of the mouth of the Piave, a few miles from Lido di Jésolo, and across the lagoon from Venice. The sun had won the last skirmish, the soft sea mist was everywhere retreating, and the orange-red roofs of Venice lay like a great, rumpled quilt over the lagoon.

A freighter carefully made its way out to sea. We sampled the sand on the beach. We took pictures of stylolites in gigantic limestone blocks along the jetty, the sort of museum-quality stylolites one never sees in outcrops but often sees in the bathroom walls of libraries.

We sat on the sea wall and ate a sack lunch — dry tuna, yogurt, green pears, and cookies. I tossed tuna chunks to a white cat hiding below among the gray limestone blocks. He was weak and frightened like the animals on the wharf at Portovenere and on the streets of Milan.

A few feet away five Spaniards took turns fishing with three poles. They never cast far, and caught nothing but big lumps of seaweed while we were there, though they had fine poles and seemed to know how to fish. Halfway to the end of the jetty an old man and woman fished.

McBride began to stare at the limestone blocks, then jumped down to look at one with his hand lens. I sat in the sun, looking toward Venice.

When I think of Venice, no matter how distant or near, I think first of St. Mark's Square and of sitting at dusk on the terrace of Florian's, writing to a special person as I imagined Byron might have written if he had begun school in a one-room schoolhouse in northern Wyoming. At Florian's a thin blue cat crept under the little table, perhaps to avoid the diving pigeons. I gave it a sugar cookie and all that remained of some expensive chocolate ice cream.

> The stratigraphy and structure of mountain belts may preserve a record of the geologic history of plate margins. However, some Mediterranean orogenic belts may have formed by uplift and gravity tectonics alone; they may not mark plate margins. For example, it is difficult to explain the structure of the Apennines by the subduction process. The partly chaotic allochthon of the Apennines seems to be the product of submarine gravity sliding and not of deformation in a trench.
>
> — W. Alvarez, 1973

Miracles still occur in L'Aquila.

The town sprang up by a miracle — 99 rioni surrounding 99 castles, 99 squares, 99 fountains, and 99 churches. The Fountain of the 99 Conduits still exists. In the evening a bell in the old tower in the Law Courts tolls 99 times — each one, deafening.

We came to L'Aquilo in the late afternoon and chose the wrong street on which to approach the center. Soon our car was surrounded by people in the one-lane street; our car was something like a boulder inching its way downslope, but upstream, into L'Aquila, as a flood of people rushed toward the inner suburbs. Pedestrians brushed against the sides of the car and, it seemed to me, looked at us menacingly. It seemed hours before we were able to park and walk on to the Albergo Italia.

The Italy Hotel, once a grand hotel in the center of the city, had declined. The rent was $18 a night for two in a room, said the clerk. Many of the tenants were very old. The ceiling of our room held a single small light bulb high above the floor. There was no light bulb in the lamp between the beds. We went out to buy one. After we had eaten, and walked to the highest point in town to see the castle — a fine example of military architecture — and sampled the ice cream, we walked back to the hotel with our light bulb. In a sitting room off the lobby I opened Eugenio Montale's little book of pieces, *Poet in Our Time*:

> *Ultimately the true artist is not the poet or the composer of aleatory music, but the man who glances at a page of advertisements or who hears noises in the street and performs the selective gesture of isolating, in that chaos, one moment or one detail which might provide a quiver of vital emotion.*

That afternoon, inching down the street amid the noise and chaos, my most vital emotion was fear.

We got up early for breakfast at a bar near the hotel. Again I had cookies with orange juice and cappuccino. At almost any bar McBride enthusiastically eats a couple of hard rolls or jelly brioches. I prefer Italian cookies, especially lemon.

I walked across the square, empty and still and cool before the Saturday morning rush, to buy some cards and a newspaper. We were the only customers sitting outside. I wrote a few lines on a card of the Luminous Fountain — with the high snow-capped Apennines in the background — to my younger son. I sent Castle at Night to my older daughter.

There is a well-marked road from L'Aquila to Pescasseroli, but we couldn't seem to find it. While we searched we saw many of the gold-tinted ancient houses that distinguish L'Aquila. When we began to pass small herds of sheep tended by black dogs, I decided we were on the right road.

A charmed snake made its way across the road. There were no other cars; we slowed for the snake's crossing.

We went into Pescina, where Ignazio Silone was born in May 1900, the month and year of my father's birth. Silone's real name was Secondo Tranquilli, but he came to hate Secondo. In one dialect it means a jail warden.

Pescina was the setting of Silone's first novel, *Fontamara*, but the lovely name does not suggest the reality of that poverty-stricken medieval village under the Fascists. Pescina is the scene of most of Silone's work, including *Bread and Wine*.

It began to rain lightly, amounting finally to little more than a mist frosting the mountain flowers in the little rocky glens. What little water struck the ground disappeared at once and went into cracks and joints. We saw no Fontamaran springs:

> *A poor, thin spring rises from beneath a heap of stones at the entrance to Fontamara and forms a dirty pool. A few paces away the water burrows into the stony soil and disappears, to reappear later in the form of a more abundant stream at the bottom of the hill.*

Our friend Roberto Colacicchi, Professor at the University of Perugia, carbonate petrologist, geological statesman in Italy, and generous and discerning host in Perugia, was one of eight geologists who collaborated on the geological map of the Abruzzi National Park. It is a lot of map and costs only about $5.

Close to one o'clock we drove into Pescasseroli, birthplace of the philosopher Benedetto Croce; it is also center for the Abruzzi National Park. It is a small town in a basin flanked on either side by beech and fir woods. Looking for somewhere to lunch, we picked the Ristorante il Caminetto, a restaurant where we discovered too late that one is wise to be careful what he samples, because the manager ran from table to table offering various dishes, and those we sampled not only remained on our table but showed up on our bill.

On a sunny Sunday morning, we went to the ruins at Pompeii. We had breakfast in our hotel room — a red orange, a little tuna on hard bread, a couple of lemon cookies, half of a hard pear — then walked to the ruins well ahead of other tourists, and managed to maintain our lead all morning. By early afternoon we were overtaken and overrun at the amphitheater where a young Japanese man photographed McBride as he peered at a detail of the structure with his hand lens. A geologist's world swings wildly from mountains tens of miles in the distance to feldspar crystals at his fingertips magnified ten times by a hand lens, to photographs of algae in oil shale 5000 times enlarged. "Things as they are" may refer to things of vastly different scales.

Pompeii and Herculaneum and Stabiae were buried under ash over 2 days by the eruption of Vesuvius in August of 79 A.D. There was a strong tremor that morning, and 3 feet of cinder soon covered the ground. Apparently the eruption was quiet enough at the start for many people to flee, but those who took cover indoors, probably to save possessions, were overcome because of darkness and a second rain of ash and cinders and lava flows as the volcano became more violent. The town was finally buried under 20–23 feet of ash and lava. About 16,000 people died, many clutching bags of coins and jewels. Pompeii was completely forgotten until it was rediscovered in 1595.

Japanese visitors outnumbered Americans. The Japanese, like everyone else, looked longest at the death scenes.

For me the lingering presence of people in the streets and squares and stores and dwellings was the most arresting quality of the ruins. Because of their long burial, they seem to be there still, unliberated by the excavators. Not all of Pompeii has been dug out and some parts will never be, though excavation at Pompeii has been less difficult than at Herculaneum.

The streets of Pompeii are straight and intersect at right angles. They are sunk between raised pavements, interrupted at intervals by stone blocks which allow one to cross without stepping off the pavement, a welcome convenience when the streets become channels at flood times. Between the stepping stones there is room for chariot wheels.

We joined a short line of people in a sidestreet waiting to tour a former house of prostitution. The rooms were small, the halls narrow, and the wall art unappealing. The cries of pain and pleasure must have been audible throughout.

A little more than 200 years ago on March 2 of 1787, J.W. Goethe started up Vesuvius. He had taken only 50 steps when the smoke became so thick he could hardly see his shoes. He pressed a handkerchief over his mouth, but it didn't help. His footing became unsteady "on the little lava chunks which the eruption had discharged." His guide disappeared. He thought it better to turn back and wait for a day with fewer clouds and less smoke. Four days later Goethe, the German painter Tischbein, and two guides — "one elderly, one youngish, but both competent men" — climbed the cone and reached the crater mouth in an interval between two small eruptions. They could not see anything because of clouds of steam from thousands of fissures. Standing at the "sharp edge of the monstrous abyss," they were suddenly shocked as thunder shook the mountain and a terrific charge flew past them. The little party retreated, covered with ash. Happy to survive, they withdrew to the foot of the cone.

McBride and I and assorted young Italians, several grandmothers, and a group of German teenage students climbed Vesuvius on Sunday afternoon. It was a brilliant day, with a few high clouds over the Bay of Naples but none over the volcano. Down the slope we could see Pompeii. Down into the crater, after a short wait, we could see tiny puffs of steam rising. There was no thunder. There was little steam. Our survival had seemed more uncertain on the way up the volcano's flank when a green Fiat cut across our path on a switchback and careened perilously on up the road.

Near the top of Vesuvius is a little shop that sells post cards, small banners, candy, and warm soda. From the children running the shop, I bought a bottle of something called "Brillante." It was bitter. I drank little of the Brillante.

McBride found a card showing a Vesuvian eruption as a yellow nuclear cloud, with successive jets of sulfurous steam rising to great heights over the summit of the stratovolcano. I looked at the faded cards in a rickety metal rack, but could see nothing interesting or attractive. I settled for a photograph of the last great eruption, which destroyed Massa and San Sebastian in 1944. Vesuvius has been quiet since, and is perhaps long overdue, if one believes in its periodicity. Certainly, in Vesuvius, volcanologists have the best long-term record of any volcano.

On the way to Tivoli, east of Rome above the Roman Compagna, I began reading *Sicilian Uncles* by Leonardo Sciascia. I have not visited Sicily. I am many books ahead of where I want to go.

Monday morning we sat on the porch outside the restaurant of the Motel River near Tivoli. I struggled with the Italian sports pages — pink sheets filled with soccer and motor racing. I learned that Michael Chang, diminutive baseliner and new teenage tennis phenomenon, had won the French Open at 17, the first American to win the French Open since Tony Trabert. The Detroit Pistons — sometimes called pistols in Italian newspapers — had beaten the Los Angeles Lakers for the third straight time in the National Basketball Association Championships.

I turn to the sport pages first, and, on some days, only. Seldom is there anything as interesting in the rest of the newspaper. Someone has won, someone has lost. The story is as good either way. Today's losers may win tomorrow, which is seldom the case in real life.

McBride does not understand my devotion to sports. He is uneasy with it. He played a little tennis in high school, but nothing else, and his coach forgot to give him his letter award. He believes that all professional sports are fixed, perhaps college sports too. He discounts Magic Johnson's injury. He believes the Lakers will win three games in the next few days merely to bring about a seventh game. I am at the other pole. I would have been one of those who at first did not believe the Black Sox scandal: "Say it isn't so, Shoeless Joe!"

It was nearly nine o'clock when we checked out of the Motel River and drove to the Villa d'Este to see the fountains. We forgot our passports at the Motel River and had to drive back later to pick them up. On a Monday in April more than 400 years ago, Montaigne came from Rome to dine at Tivoli. He wanted to see the famous palace and garden, which were incomplete and "not being continued by the present cardinal." He was struck by the sights and sounds of the fountains — the noise of cannon shots, the noise of harquebus shots, and the songs of birds. He wandered through the "gushing of an infinity of jets of water" and by the head of pillars where "water comes out with great force, not upward but toward the pond."

I am also partial to waterfalls and fountains. The thing that first made them special for me was a trip to Yellowstone National Park with my parents, brother, and sister when I was a boy. We saw Grand Geyser and Old Faithful Geyser. And I tried to understand geyser action, the heating of water in the pipes, the rise in temperature toward the boiling point, the quiet flow of fountaining water a few feet above the vent, the lowering of the boiling point in the pipes, and the explosive change to steam and violent eruption. I remember the feel of the sinter around the geysers and the tremble of the earth as Old Faithful erupted. I was delighted when a change in the wind direction blew spray on me. I have a dim memory of my mother later pulling out a sack of oatmeal cookies she had made to bring along on the trip.

We drove through rain deep into the center of Sulmona, looking for a hotel in that medieval city. It was cold. The rain came in bursts and gusted along the meandering narrow streets off of the piazza. The gray sky and threatening storm brought back memories of days and evenings when I tended the drop herd on Gooseberry Creek near Worland, the new lambs frightened and wet and bawling as the rain became sleet blowing against red-brown and green badlands.

I saw a small hotel sign in the middle of the scarred street. A middle-aged woman with light blue-green hair said we could not stay where we had parked. I watched the car while McBride went to rent a room. A young man came out of the hotel, the Stella, to find us a parking place. He looked up and down the street, and walked over to a car parked near a municipal building. He recognized the car. He found the owner and got him to move it, and we pulled into a spot beside a dumpster in front of a church.

We walked through the center of Sulmona. Small shops there featured sugar-coated almonds on tiny trees — dark red, orange-pink, olive-brown, brilliant green, and blue-green leaves and fruit. I bought a pocket knife for my younger son and later sold it to him for a dime — insurance against cutting off our friendship. I bought a picture for a writer friend, a scene of mountains and a valley made with metal pieces.

For myself, I bought a cotton tennis shirt with an insignia that resembled a spotted star thistle — many small heads of pink-purple flowers and narrow leaves. The cotton was dyed pale olive, but the minute I got home and put it on it began to change color. Dusky yellow spots appeared down the left side below the thistle. I started to eat a peach and the juice dripped on the shirt, which broke out in moderate-yellow and yellow-gray splotches, resembling the first stage of some rare tropical disease. Before I washed the shirt, pale green-yellow piping began to appear. It never recovered its original color and became whatever dull olive or yellow or brown the dye changed to after each washing or rain.

"In this town everything closes early," said the manager of Trattoria da Ginos. No doubt this is correct, but it is true for all the small towns and cities in the Apennines.

"In 1933 an earthquake wrecked the church of San Francesco della Scarpa," said McBride. We walked through the thin dark streets of Sulmona, the birthplace of Ovid. The rain had stopped and the streets were empty. That Ovid preferred Rome to Sulmona is not surprising.

> *New structural and geochronologic studies provide strong support that crustal shortening due to plate convergence is the probable cause of much of Northern Apennines orogeny. . . . These studies indicate that crustal shortening associated with continental collision is not confined to the suture zone (the former plate boundary) but is heterogeneously distributed throughout a broad zone extending for considerable distances into the continental crust of the collided plate.*
>
> — Roy Kligfield, 1979

"Robert L. Folk, the sedimentary petrologist, may be stopping in Viterbo with students," said McBride. Given the route we were taking down the Apennines, then north again, Viterbo was roughly on our way to Gubbio. We decided to look for Bob Folk. We would not have visited Viterbo otherwise.

Bob travels by train in Italy and stays at the least expensive hotels. We inquired at several on the Via della Cava, a frighteningly crowded street. He was not registered. We decided to stay at the Albergo Roma, the most Folkian hotel we had seen. McBride described Folk to the young woman at the desk. She said she had not seen anyone like that for a long time, but she would tell us if he checked in.

We walked to a small square for ice cream. The same people passed our table many times — middle-aged men walking arm in arm, young men laughing and shouting, a young woman pushing a stroller containing a fat baby who slept through the noise. The people of Viterbo did not appear prosperous. The square and streets were dirty. I later read in Montaigne's journal that he did not find a single gentleman among the inhabitants of this town in 1581 — "they are all laborers and merchants."

From an extravagant menu of desserts I chose a banana split. As pictured, it was beautiful. I thought of Tommy Schultz's drugstore in Worland and Saturday afternoons I spent there after I collected from my paper route. Tommy Schultz sold a caramel malt and a grilled-cheese sandwich for 55 cents in 1941. The beautiful-appearing banana split in Viterbo sold for 6000 lira, a little less than $5.00.

Italians are inarguably marvelous cooks. But they have yet to learn how to delectably combine fruit, syrup, and ice cream in a dish.

The dinner wine: Est! Est!! Est!!!

In the morning we walked to the railway station. There were few hotels. Beyond the station was a three-star hotel, but neither McBride nor Folk would stay there. I said Bob might — he had some money for research expenses. In the lobby three fashionably dressed young men in blue suits and red silk ties quickly surrounded us. They had no record of a Folk. They had not seen anyone who resembled him, certainly not lately, said one in rapid Italian. They did not really listen to McBride's description of Folk. They walked with us through the lobby to the street.

After dinner I stopped in the TV room at our hotel. I began to watch a gangster melodrama — dead men and live men that looked dead tossed by gangsters and lawmen into car trunks in the middle of the night.

The TV room was the popular room of the hotel, and no one stayed in the adjoining bar for long. Perhaps it was more than a TV room. Striking young women met and left frequently with middle-aged men. The young women seemed to live at the hotel. One was waiting for her friend Jessica, and every few minutes she walked into the lobby and called loudly for Jessica. Jessica had the look of one who had been influenced by Claudia Cardinale. They left together. Another young woman came in carrying Camillo, a small gray monkey wearing green pants. Camillo talked, sucked the young woman's fingers, and swung on the frayed curtains. He picked up a cigarette butt from behind the dilapidated couch and chewed on that. When I told McBride about Camillo, he listened, but said nothing.

Approaching Gubbio in the early afternoon — after stopping in Perugia to buy one of the fabled roast chickens, whose spicy flavor McBride has been unable to duplicate in Austin, Texas — we pulled off for lunch on a low hill incarnadine with blooming poppies. The narrow dirt road followed a high contour on the hill; the poppies were neatly strung in an unbroken arc below the road ending at a red-roofed farmhouse. The cosmic collision hypothesis, suggested by concentrations of iridium 100 times or more the crustal averages in claystone at or near the Cretaceous-Cenozoic boundary near Gubbio and other places, was a hypothesis hard to hold in the mind while sitting in the sun, the brilliant flowers streaming down the hill to a stripe of yellow earth. Though asteroids collide with the earth every 40–50 million years, cosmic catastrophes seem out of the question in Umbria.

We stopped a few minutes at the gorge where the first iridium anomaly was found. I stood with one foot on the Cretaceous, the other on the Paleocene, earliest epoch of the Cenozoic. I sampled the red-brown claystone at the boundary, and McBride took my picture as I gazed upsection toward Recent rock. If a large asteroid body — 6 to 9 miles in diameter — had hit the earth 66 million years ago as Alvarez, Alvarez, Asaro, and Michel have postulated, the extinct dinosaur lines and those of many lesser creatures ended cataclysmically somewhere stratigraphically below my left little toe and above an abandoned brown sweat shirt on the bare hogback. The hypothesis: The asteroid collision threw up a gigantic dust cloud of asteroid and earth materials, the sun was obscured for months, photosynthesis stopped, the earth was chilled, and the dozen or so dinosaur genera still living — out of about 240 genera that once existed in the Mesozoic — died. The last dinosaurs lived chiefly around the drying western interior seaway of North America.

When the cosmic collision hypothesis was first proposed to explain the demise of the dinosaurs, it appealed to a variety of minds. There are hundreds of hypotheses to account for that particular mass extinction, but this one is certainly favored by journalists.

Paleontologists were particularly skeptical of the cosmic collision, arguing that the Late Cretaceous extinctions were not catastrophic. Only a small percentage of all dinosaur genera were still alive in the latest Cretaceous. Ammonites, the best guide fossils for worldwide correlation, did not become extinct suddenly at the end of the Cretaceous. They merely became less diverse. In the middle Late Cretaceous there were about 100 genera of ammonites. By latest Cretaceous, only 10 genera were left.

The climatic cooling following the hypothetical cometary collision spared insects, birds, mammals, lizards, turtles, crocodiles, burrowing snakes, and the majority of plants in low and middle latitudes. "Whatever the cause of their demise, the roughly simultaneous and abrupt death of the phytoplankton that devastated the marine nutrient pyramid seems more than enough to account for other marine extinctions," said Preston Cloud.

We stopped overnight at Bagno di Romagna on the way from Gubbio, spending the day in the field at localities of Contessa claystone. We were curious to see *Trichichnus* burrows. We wondered how deeply the burrow-maker had marked the claystone, an exceptionally thick (45 feet) turbidite, termed a megaturbidite by the great submarine-fan authority, Franco Ricci Lucchi, Professor at the University of Bologna. Deposition of such thick fine-grained beds should have wiped out all local benthonic animals.

We wondered if this had happened, and how quickly the area would be repopulated if it had been, and by what?

We had little trouble finding the Contessa localities, but much trouble in placing them on our map. The narrow Apennine roads sometimes defy location. Italians also remove road signs, said McBride.

We forded a shallow river to reach one locality. There was a remarkable amount of trash along the banks and on the flood-plain — unmatched sneakers, plastic bottles, torn and weathered pants and panties. On the stream bed in mid-channel bars, there were football-size boulders, fringed by algae around the smooth bald pates of polished sandstone. Where a thin film of mud and algae covered the worn boulder tops, the rocks were slippery and treacherous to step on.

I cut my wrist breaking through thick brush above the flood-plain. Talus covered the steep mountain apron and presented an unpassable route upward to the Contessa. McBride worked his way out a few yards onto the apron but turned back. I did not make it that far, falling behind in my little war with the brush, slowed down by constantly looking out for poisonous snakes. McBride never sees snakes. But I see them everywhere in the field: every swaying grass clump and rattling thistle makes me jump.

The Contessa megaturbidite here was laid down by deep ocean currents flowing north. All the other sediment, according to our reading of the current directions from flute casts on the soles of the sandstone, came from the north. We found a great variety of trace fossils in sandstone and claystone below the Contessa, but none in the Contessa sandstone. *Chondrites* occurs in the fine-grained Contessa, down to about 15 inches below the base of the hemipelagic bed, and *Trichichnus* to about 55 inches. A tourist bureau might bring people here to see the high smooth green hills, broken badly during mountain-making, and the narrow country roads like giant spider webs dropped on an unsuspecting elephant herd, but tourist bureaus seldom leave beaten tracks. In a light rain that came and left and came again, the beauty was in the soft green hills, the isolated belts of brown and gray rock, and the still sky.

Modena is a pleasant place but is the most expensive one we stay in regularly. We approached it looking for the enchanted Farnese tower, the slender one built on the broader one depicted in Stendhal's novel *The Charterhouse of Parma*. But the tower never existed in either Modena or Parma.

Daniela Fontana and Rodolfo Gelmini, of the University of Modena, worked with us on the metamorphosed Triassic Verrucano sandstone in Tuscany. The four of us stayed for a few days at Albinia, a small town that has not made its way into guidebooks or into anyone's printed journal. With Gelmini driving and shouting at slower cars and people, we roared down the Tuscany coast to Ansedonia where Puccini may or may not have composed part of *Tosca*. From the old wall, once we careened our way up to it, we could look across the lagoon to Monte Argentario where the Verrucano crops out. McBride began to choke up and lose his voice. Aromatic plants — not the rocks — had overcome him.

The Verrucano sandstone is sufficiently metamorphosed to be perplexing to sedimentary petrologists — our specialty — but too little metamorphosed to interest metamorphic petrologists. The Verrucano is a great sandstone sequence with few advocates.

Last fall a delegation of Modena business men came to Salt Lake City for a visit, a sort of Chamber-of-Commerce tour. I was surprised to learn that Modena is a sister city to Salt Lake. The business men were studying our way of doing things. I love Salt Lake as if I were a native. I love the four seasons, the deer eating my tulips, lazuli bunting and green-tailed towhee feeding from a pie pan on my deck, the lights at night, the lake, the broken mountains, a pale brown-haired woman. But I cannot imagine that Modena has a lot to learn from us.

The Italians were small, trim, stylishly attired, and affable on television. They did not wince when a local announcer called their city Mow-deena.

Though McBride has told me many times, I cannot recall the kind of cars made in Modena. Maserati? Or perhaps it is Ferrari, or both. Though Modena is a large commercial center in Emilia-Romagna, it does not look like a city with car factories. None of our friends in the geology department are experts on the business life. And none of them seems to know that Mary of Modena, unlucky consort of James II of England, was born there. But they do know where to find a good restaurant and fine Lambrusco and the best sandstone outcrops in the Apennines.

We stayed the night in Fontanelice, our next-to-last night in Italy. We came to take another look at *Trichichnus*, one of the smallest megascopic trace fossils. On a light olive-gray claystone face *Trichichnus* resembled short burnt-gold hairs unaccountably standing on end, but poised to vanish when the sun struck the shadowed wall.

McBride has a keen eye for the tiny trace fossil. He first saw this one in basin-plain and outer fan-lobe strata of the Marnoso-arenacea Formation when we stopped in the canyon on the way to Castel del Rio, a town devastated in World War II. I was crawling along looking at burrows of a trace fossil ten times as wide as *Trichichnus*. McBride did not shout, though almost any paleontologist would have hollered.

The *Trichichnus* animal burrowed vertically down through silt and clay and then horizontally on top of sand. It is most abundant in siltstone and claystone.

We ate dinner in Fontanelice down by the river below the channel sandstone of the Marnoso-arenacea. The cafe offers a small menu, the river setting, dusk coming down the canyon wall, and low prices. A man on a white gelding, and his daughter astride a sleek bay, galloped up to the porch, tied the horses to a tree, and had a glass of wine. Refreshed, they rode off, shouting and whipping the horses into a gallop on the graveled road. Our colleague and friend Giani Zuffa — incoming president of the International Association of Sedimentologists — said he did not know them or where they came from.

Back at Giani's house I mentioned how much I admired Lawrence's *Sea and Sardinia* and how much I believed Lawrence had learned about the people in such a short time. Francesco and Laura Dessi, friends of the Zuffas from Florence, began to speak excitedly. Francesco's father, Guiseppe, it seemed, had written the fine novel *Paese D'Ombre*. *Paese D'Ombre* evokes best, said Francesco, the people and the land of Sardinia. I promised to read the book.

"I was here informed of a stupid thing I had done in forgetting to visit a mountaintop ten miles this side of Loiano and two miles off the road, from which, in rainy or stormy weather and by night, you can see flames coming out to a very great height; and my informant told me that in the big upheavals there are sometimes disgorged little pieces of money with some figure or other," said Montaigne.

McBride and I also did a stupid thing at Loiano. Two years ago we stopped there to sample concretions, part of our long-term study of carbonate cementation. From the maps we carried, or from the articles on the several formations, or from lack of perspicacity, or from some combination of these things, we sampled the wrong formation. Last year on our way north we stopped again and had an ice cream cone and found and sampled the rocks we had intended. We did not see flames or little pieces of money, nor did we when we were there before.

The hotels in Loiano had a poor reputation in Montaigne's time: "There are only two hostelries in this village, which are famous among all those in Italy for the treachery that is practiced on travelers in feeding them with fine promises of every sort of comfort before they set foot to the ground, and laughing at them when they have them at their mercy: about which there are popular proverbs." With that warning in mind we did not stay overnight. Instead we headed north to Pianoro, where we measured the dimensions of concretions and took pieces for isotope and petrographic study. From previous samples we found that the quartz has been badly fractured during earthquakes and the mountain building that formed the northern Apennines. This is a complication. The original characteristics of the quartz and other minerals in the sandstone may not be determinable.

After Pianoro we were caught in a phantasmagoria of weekend traffic — people on their way back to Bologna on Sunday evening. At Bologna — still under the spell of Loiano — we missed the turn northwest onto the A1 to Salsomaggiore Termi.

Our delay in traffic and at Bologna meant we could not reach Milano, which I believe Earle was not keen to do anyway. He does not share Bob Folk's or Italo Calvino's high opinion of Milano and passes on half an excuse. Milan traffic or air pollution may suffice.

Salsomaggiore Termi was a welcome surprise. It was not in any of our books. It provides a watering place for those temporarily fleeing Milan and, in the inexpensive little hotels, a place for widows and aunts seeking resort gaiety at a low cost. We reached the town a little after 7:00 p.m. It was 8:00 before we found a hotel, the Albergo Venezia, and then we arrived by its back door. It took 10 minutes for the proprietor to find us as we knocked and rang the bell.

We took a ritual stroll along the piazza, sampled the ice cream, which was above the low Loiano standard, and discussed the wisdom of field studies in geology. We began to plan the next year's field work in Italy, certain we would be there again, unable to leave the broken rocks alone.

The Sea at Portovenere

The sea at Portovenere swells and moves in green shadows below wave crests and pales to shell-white where it breaks free of its orbits to collide with boats. Crests break off and the water moves faster than the wave-form. Along the dock the renegade water rocks a small boat, but the boat's only occupant, a gray cat — tipped black on its ears and down its spine and tail — rides the water, too weak to leave the boat. A female cat — similar enough to the boat cat in color and head shape to be its sister — sits by my feet in the shadow of a bench on the carbonate promontory.

Above me stands the walled church of St. Peter, a Genoese building with marble paving laid in the sixth century. The church and the stone walls rise from Triassic rock 200 million years old that dips steeply toward the bay, is possibly overturned, and is certainly folded and faulted all along the outcrop. The sea attacks the faulted beds, shaping the rock as it has done through countless centuries. The great Moore figures of mothers and children change and move inland — hole by hole, crystal by crystal.

Ancient houses, varicolored and salt-sprayed to the highest stories, line the port and the main street of Portovenere. The pastel colors of the buildings are bright when the sun strikes them — pale and yellowish orange, moderate and grayish pink, pale and light red, yellow brown, and yellow green. The paint covers cracks partly filled with centuries worth of silt, salt, and carbon. Looking like Christmas boxes, the buildings bob in the changing light, cloud-shadowed, on a perpetual early Sunday morning.

The Genoese, who considered the site strategically important, fortified some of these buildings, all of which sit like unstable crustal blocks on an unsteady terrain, dependent on the support of common walls to remain upright and isostatically in balance. If one moves, so might they all, perhaps catastrophically.

I seem to have come here to watch the waves and not to look at the rocks. Capricious behavior for a geologist. But the struggle between land and sea has always attracted throngs of watchers. The sea attacks even the strongest rock, and we are drawn to it. Here the rock is called Portoro — limestone and dolomite, jet black with white veinlets and gold stylolites. Ornamental stone.

Long ago the sea took away the loose grains and talus along the cliffs as well as those grains and crystals held by weak cement. Waves hurl themselves against the land, expending energy collected offshore in the Ligurian Sea. Water flows into holes and fractures, forces its way into hidden cracks, and runs along corridors of caves, its thunder rolling from rock walls that have risen above the sea during the last several million years. The wave-cut cliffs have receded grudgingly. The shore trades a narrow high cliff here for a new low beach there. The land straightens where it can, curves where it must, and presents the least profile to the sea's ceaseless attack. The shoreline is in equilibrium with the energy expended on it.

Those parts of me that ache are soothed. I find relief by the sea. Staring at the waves, I am carried up green-shadowed crests and down into dark-blue troughs, a traveler filling hollows of fast-running floors.

The cat lies quietly at my feet. None of the cats I can see along the dock have moved. They are starving, sifting away day by day, cut and infected and scarred in fights over cafe scraps.

A few hundred yards to the west, on the other side of the promontory, tourists and local people from La Spezia swim and sunbathe in Byron's Grotto. In the beauty of the sea caves and rock-arches, and in the creek surrounded by stratified rocks, Byron found the inspiration for his poem *The Corsair*. He would also have sought and found ready acceptance among the perfumed women up from Rome.

Yesterday, the waves threw boulders against the shoreline cliffs. In the Saturday storm, no one could swim in Byron's Grotto, and only a few came to watch the sea catapult pebbles and cobbles onto the carved rock of the narrow wave-cut terrace. The Genoese church trembled, waves broke over the steep path to the church, and the carbonate promontory shook as the water struck the rock. The Portovenere cats crouched in the boats along the dock, riding the storm out. The shoreline moved imperceptibly back.

I think of how little of my life I have actually spent by a sea. I regret that this secret feeling for the sea, this feeling that so many have, came too late.

I pick up the elemental sound of the ocean against the cliffs. The swelling sea falls. From the remains of lunch, I share a hard roll from the Hotel Belvedere with the gray cat. When the scraps are gone, she purrs faintly and offers her stomach to be petted. The sea moves sediment in drifts down the shore and north between the promontory and Isola Palmaria. In Portovenere they say Byron swam in the Gulf of La Spezia from Portovenere to Lerici, 5 miles to the northeast.

Byron was capable of extraordinary things. Perhaps he did swim to Lerici. Perhaps he could have swum on to Viareggio.

Finishing Strong

Nature is the source from which, after all, we get all we know about our feelings, or what we flatter ourselves are our feelings. You can't explain these inchoate yearnings and fears inside you except with reference to the world outside. Those relationships must have been old and taken for granted when Homer was a kid.

— Howard Nemerov, 1988

On a chilly spring morning in 1979, my father and I left the corral at the 2-B Ranch and rode north along Bridger Creek in north central Wyoming to round up a few head of stray cattle and brand the calves.

He was 79, and recovering from a near fatal heart attack. His heart had stopped on the way to the hospital and it hadn't started again until the head nurse hit him 19 times with a cardiostimulator.

Who would ask him not to ride and help in the roundup? His brothers, one older and two younger, wouldn't have thought to say anything — they would have done the same thing. My sister and brother and I knew better than to say anything though my sister couldn't resist. "You be careful now, Daddy," she said.

My father and his brothers and the other mountain men and women took care of their own, doctored their animals and children, and lived close to the land. Being born in these mountains and riding horses on the badlands of the high basin floors, as well as riding through the talus of the mountain's apron and up its faulted flanks to gather cows and sheep near the mountain peaks, was a wondrous way to live — one that is almost gone now.

There was hardly a rock or trail or spring within a hundred square miles that my father did not know intimately. He was a tireless rider. He often left the cabin at sunup and rode all day, taking a water bag but seldom any food.

He saw everything. And, when I rode with him, he told me what he saw.

A few weeks ago my feelings about that last roundup with my father got so strong and persistent, I retraced our route on a geological map. We crossed Bridger Creek that spring morning and stopped in the deepest part of a meandering reach to let the horses drink. I recall the images in detail because they were just as they had been at that time of year since I was a boy. The horses had a morning blow; their breath formed white geyser clouds that lingered in the chill air. We got off on the sandy bank. The low hills were sharply shadowed in the early light. The night had moved away, west of us by then, possibly covering bands of rock slapped on the Pacific border in a sloppy fit of mountain making.

On our way on up the valley we passed holes where rodents had burrowed into the alluvium and the red soil of an alfalfa field. They watched us from low mounds, their clustered homes dug out of hematite-stained clay and silt, grains that had been carried back and forth across a tidal plain bordering the Early Triassic sea some 240 million years ago. "A horse could break a leg in one of those holes," said my father. He had warned me about the holes for 50 years.

Details of the landscape that reminded him of old times, old sorrows, or snowstorms or roundups frequently got him reminiscing out loud, as when we passed the worn butte where his sister Marion was buried.

To the north, Sohio Petroleum had drilled a well to 1133 feet. My father remembered they left a gate open and some of the cows got out. But everyone up and down Bridger Creek, my uncles especially, wanted to see a gusher shoot up. They did find oil had stained and filled some pores in Phosphoria dolomite and Tensleep sandstone. "There was too little oil," said Sohio. They were there 3 weeks and drove away.

After I became a geologist, I gave my uncles a lot of free advice over huge breakfasts of sourdough pancakes and bacon. I warned them about investing in wildcat wells or anybody else's play. My father didn't have any money to invest, and he was even more skeptical of hitting a bonanza than they were. They all spoke wryly of finding oil, and all but my father had received some money through the years from leases. They broached the subject of oil warily, but once they started talking about it they couldn't quit, and would go on and on questioning me about oil formations. Uncle Dave often asked, with a wicked grin, why I couldn't find oil for them.

We had crossed the axes of several westward-plunging anticlines, low gentle arches of Triassic rock, the reason the explorers drilled for oil. As we rode along, the color of the buttes turned from dark maroon to reddish brown in the wedge-shaped shadows. They looked like weathered bricks. The red siltstone was inclined a few degrees to the northwest and north.

High above us a sheepherder had built a pyramid of siltstone, ripple-marked and mud-cracked. The sound of a bird singing made me want to write a poem or break into a gallop.

We turned sharply west and climbed to an escarpment held up by thin olive-gray limestone. "This is a place to watch for rattlesnakes," said my father.

He always sat his horse a little off center, his weight to the left, his hat low, shadowing a wide face, weather lined and red. He weighed no more than 150 pounds then, but still seemed a much larger man. I followed him up the narrow trail through the limestone outcrop. Slowly, majestically, the beautiful morning went by. We passed a bench of yellow sandstone, the cross-strata suggesting deposition in an ancient tidal channel; a sunning rattlesnake suggested we take a higher trail.

He saw the small bunch of cows and calves before I did. He knew beforehand where they would be. They were a long way off, a rebel band, far from the main herds in the alfalfa pastures. He told me that wolves once ranged over these hills, and there had been bear too. Now there were coyotes and deer and cows in the high isolated gullies and near the small springs that fed tiny upland meadows.

At a turn in the fence, part of the winter's deer-kill danced on the barbed wire. The fence held some carcasses upright. The others lay together, corded like stacked oak. The deer had sheltered themselves against the fence below a band of volcanic rock — airfalls from Eocene volcanoes — and the spring storm had swept over them.

We threaded through the deer to where the sandstone was covered by sheet wash. We went on in single file, my father still in front. He broke the bay into a trot, headed on a diagonal across the gently sloping green claystone.

We swung in back of the strays to drive them home, and they turned obediently downslope, their rebellious wandering quelled for that day.

On a high flat we cut out a neighbor's heifers and a steer, and sent them westward. My horse stumbled, got his feet under him, and jumped forward, nearly dumping me in the sagebrush.

A light breeze arose. We saw a small band of deer a long way off. Indian Paintbrush, their bracts and upper leaves brilliantly red, colored the ground.

Back on the escarpment above the alfalfa fields we paused to let the cows make their way slowly down the narrow trail through bluffs of red siltstone deposited in cycles. When I was a boy my father had pointed out what he saw as the face of a middle-aged man sculpted in the stone at the top of the cliff, a face released from the siltstone, he said, by running water and wind, a face I later thought resembled one of Rembrandt's last self-portraits. I had seen other faces there when I was a boy, and figures that were frightening. Once at dusk I fancied I saw a tiger, striped red and brown, lying above the trail, its forepaws crossed.

We sat in silence until my father, choosing a moment he seemed to have waited for, turned in his saddle and said, "It just wasn't my time to die."

I didn't know what to say. He had never spoken to me that way before.

"You'll live to be a hundred," I said.

He laughed. "I wouldn't want to."

We turned down the trail, the orange dust falling slowly in the light and settling on us and on the cows and calves.

Following the heart attack, he described the disoriented feeling he had on the way to the hospital and how surprised he had been at how sharp the pain was; he also described how, in the hospital, he had felt as if he were hovering near the ceiling of his room and looking down at himself on the bed. He said that he hadn't cared, then, if he died. But he was happy enough to be living now, he added.

A couple of hours later we pushed the little band of cows into the big herd gathered in the lower pasture near the corral. We joined my brother and two uncles and the other riders and crowded the cattle through the corral gates.

It took several hours for the herd to settle down and for all the calves and mothers to find each other. The bawling finally stopped. We went into the corral and began roping the calves, dragging them to an open place, wrestling them down, then vaccinating, cutting ears, cutting pouches of bulls, and putting the iron on them. The air was stifling. The breeze had died. Uncle Dave cut, my brother roped, and my father branded. I wrestled calves and cows out — a student of the earth getting a taste of it.

When we let the calves and cows out, they streamed toward Bridger Creek.

The men walked to the spring below the cabin to wash and drink cold beer. Dave came out of the cabin carrying a fifth of bourbon. It passed from hand to hand. No one wiped the lip of the bottle. A boy jumped smack in the mud under the spring bank. My father took a pull from the bottle and passed it to my brother. As I stood beside them in the late afternoon, scenes from scores of other roundups and brandings on the mountain came back to me.

My father died 4 years ago. I was in Portovenere, on the Tyrrhenian Sea, at the time, where I was studying sand and gravel in a short-headed, high-gradient river. A month earlier, in Worland, my father remembered the time he told me about some oil seeps in fractured and faulted granite on the east end of Copper Mountain. I doubted him so we drove there and walked up into the rounded, broken-granite knobs. I was wrong. My father was right. I watched the black oil oozing over feldspar and quartz and hornblende, down a sun-warmed granite face, and tried to think of an explanation. I was very young, officially an oil-patch geologist for only a few months. Everything I had learned in school said oil did not originate in granite or commonly reside there. But it was there anyway. That was a great lesson for a brash young geologist, one that has come back to me many times.

Bibliography

Adams, F. D. (1938), *The Birth and Development of the Geological Sciences.* Dover edition, 1954, Dover Publications, New York, 506 pp. Traces geology from Greek and Roman times, through the Middle Ages and the Renaissance, to the mid-nineteenth century.

Alvarez, Walter (1973), The application of plate tectonics to the Mediterranean region. In: *Implications of Continental Drift to the Earth Sciences*, Vol. 2, Academic Press, London, pp. 893–908.

Alvarez, L. W., Alvarez, Walter, Asaro, Frank, and Michel, H. V. (1980), Extraterrestrial cause for the Cretaceous–Tertiary extinction. *Science*, 208, 1095–1108. The original article proposing comet or asteroid impact as the cause of the Cretaceous–Tertiary mass extinction.

Alvarez, Walter, Alvarez, L. W., Asaro, Frank, and Michel, H. V. (1982), Current status of the impact theory for the terminal Cretaceous extinction. Geological Society of America, Special Paper 190, pp. 305–315.

Arthur, M. A., and Fischer, A. G. (1977), Upper Cretaceous–Paleocene magnetic stratigraphy at Gubbio, Italy. I. Lithostratigraphy and sedimentology. *Geological Society of America Bulletin*, 88, 367–371.

Barzini, Luigi (1983), *The Italians*. Atheneum, New York, 356 pp. The Italian way of life—style, customs, temperament, manners, morals. A witty book that becomes strangely tragic and melancholy. I admire the writing.

Bruhn, R. L., Picard, M. D., and Beck, S. L. (1983), Mesozoic and Early Tertiary Structure and Sedimentology of the Central Wasatch Mountains, Uinta Mountains and Uinta Basin. Utah Geological and Mineral Survey, Special Studies 59, pp. 63–105.

Bruhn, R. L., Picard, M. D., and Isby, J. S. (1986), Tectonics and Sedimentology of Uinta Arch, Western Uinta Mountains and Uinta Basin. *American Association of Petroleum Geologists*, Memoir 41, pp. 333–352.

Calvino, Italo (1983), *Marcovaldo*. Translated by William Weaver, Harcourt Brace Jovanovich, New York, 121 pp. In 20 stories set in a drab northern Italian industrial city of the 1950s and 1960s, Calvino weaves his fantastic vision.

Chamberlin, T. C., and Salisbury, R. D. (1906), *Geology: Geologic Processes and Their Results*. Henry Holt and Company, 684 pp. "Geology is essentially a history of the earth and its inhabitants," said Chamberlin and Salisbury. Wonderful illustrations. If students had such a grand book today, they might be tempted to read it.

Cloud, Preston (1988), *Oasis in Space: Earth History from the Beginning*. W. W. Norton and Company, New York, 508 pp. Stresses the early earth and Precambrian history; written for a general audience by a superb writer and geologist.

Dana, E. S. (1932), *A Textbook of Mineralogy*, fourth edition, revised and enlarged by W. E. Ford. John Wiley & Sons, New York, 851 pp. I still use my copy, which I bought in September of 1947 for $5.00.

Dickens, Charles (1846), *Pictures from Italy*. Reprinted by Ecco Press, New York, 176 pp. Begun in 1844 as a series of letters from Italy. "This book is a series of faint reflections—mere shadows in the water—of places to which the imaginations of most people are attracted in a greater or less degree, on which mine had dwelt for years, and which have some interest for all."

Ekdale, A.A., and Bromley, R. G. (1984), Sedimentology and ichnology of the Cretaceous-Tertiary boundary in Denmark: Implications for the causes of the terminal Cretaceous extinction. *Journal of Sedimentary Petrology*, 54, 681–703. They argue against "meteorite" impact as cause of the terminal Cretaceous extinction event. They favor sea-level regression, extensive volcanism, and calcite dissolution in ocean surface waters.

Faul, Henry, and Faul, Carol (1983), *It Began with a Stone*. John Wiley & Sons, New York, 270 pp. Includes an interesting account of some Italian contributions to the birth of modern geology.

Fowles, John (1969), *The French Lieutenant's Woman*. Little, Brown, and Company, Boston, 480 pp. If you liked this novel, which I did, you will want to read A. S. Byatt's *Possession*.

Goethe, Johann Wolfgang von (1982), *Italian Journey* (1786–1788). Originally entitled *Italienische Reise*, North Point Press, San Francisco, 507 pp. Translation by W. H. Auden and Elizabeth Mayer. Goethe was "hauled" up Vesuvius by a guide wearing "a stout leather thong around his waist." He grabbed the thong, "guiding his own feet with the help of a stick."

Hallam, A. (1983), *Great Geological Controversies*. Oxford University Press, Oxford, 182 pp. Considers five case histories going back to the beginning of geology.

Haycraft, John (1985), *Italian Labyrinth*. Penguin Books, New York, 314 pp. Italy in the 1980s.

Harris, E. L. (1992), *Native Stranger—A Black American's Journey into the Heart of Africa*. Simon and Schuster, New York, 315 pp. The year-long inner and outer journey of a black American along invisible threads through Africa. Impossible to put down or ignore.

Harrison, B. G. (1989), *Italian Days*. Weidenfeld and Nicolson, New York, 477 pp. A lively travel book—Milan, Venice, Florence, San Gimignano, Rome, Naples, Molise, Abruzzo, Puglia, and Calabria. A woman's view of Italy from the outside.

Hudson, W. H. (1918), *Far Away and Long Ago*. Eland Books, London, 332 pp. A boy's life and experiences growing up in Argentina—growth of an animist. The best and most interesting book by a great writer.

Judson, Sheldon, Kauffman, M. E., and Leet, L. D. (1987), *Physical Geology*, seventh edition: Prentice-Hall, Englewood Cliffs, New Jersey, 484 pp. A popular college text.

Kiersch, G. A. (1964), Vaiont Reservoir disaster. *Civil Engineering*, 34, 32–39.

Kligfield, Roy (1979), The Northern Apennines as a collisional orogen. *American Journal of Science*, 279, 676–691.

Krutch, J. W. (1958), *Grand Canyon*. Anchor Books, Doubleday and Company, Garden City, New York, 252 pp. The Grand Canyon whole—width, depth, time, values, proportions.

Lahee, F. H. (1941), *Field Geology*, fourth edition. McGraw–Hill, New York, 853 pp. A must for students when geology was a field science.

Lawrence, D. H. (1916), *Twilight in Italy*. Reprinted by Penguin Books, New York, 174 pp. "When one walks, one must travel west or south. If one turns northward or eastward it is like walking down a cul-de-sac, to the blind end."

Lawrence, D. H. (1932), *Etruscan Places*. Reprinted by Penguin Books, New York, pp. 97–215. "The Etruscans, as everyone knows, were the people who occupied the middle of Italy in early Roman days and whom the Romans, in their usual neighbourly fashion, wiped out entirely in order to make room for Rome with a very big R."

LeRoy, L. W., compiler and Editor (1950), *Subsurface Geologic Methods*. Colorado School of Mines, Golden, Colorado, 1116 pp. A very large collection of articles on the methods of subsurface geology.

Longwell, C., Knopf, A., and Flint, R. F. (1941), *Outlines of Physical Geology*, second edition. John Wiley & Sons, New York, 381 pp. Once the most popular college text for beginners.

Love, J. D., and Boyd, D. W. (1991), Pseudocoprolites in the Mowry Shale (Upper Cretaceous), northwest Wyoming. *Contributions to Geology, University of Wyoming*, pp. 139–144. Co-authored by the hero of *Rising from the Plains* and the hero of "Tracemakers".

Lowrie, W., and Alvarez, W. (1975), Paleomagnetic evidence for rotation of the Italian Peninsula. *Journal of Geophysical Research*, 80, 1579–1592.

McBride, E. F., Picard, M. D., Fontana, Daniela, and Gelmini, Rodolfo (1988), Sedimentology, petrography and provenance of the Triassic Verrucano Group, Monte Argentario (Tuscany, Italy). *Giornale di Geologia*, serie 3a, 49(2), 73-92. Rocks too hot at birth for sedimentary geologists and too cold for metamorphic geologists.

McBride, E. F., and Picard, M. D. (1991), Facies implications of *Trichichnus* and *Chondrites* in turbidites and hemipelagites, Marnoso-arenacea Formation (Miocene), northern Apennines, Italy: *Palaios*, 6, 281-290. On the origin and ecologic implications of *Trichichnus*, a thread-like burrow possibly the work of sipunculan worms.

McCrackan, W. D. (1907), *The Italian Lakes*: L. C. Page and Company, London, 362 pp. "The Italian lakes express perennial youth and freshness, joyousness and peace."

Montaigne, Michel de (1983), *Montaigne's Travel Journal*: North Point Press, San Francisco, 175 pp. In September of 1580, Montaigne set out for Rome by way of Austria and Switzerland, a journey of 17 months by a wide-awake traveler. Delightful.

Montale, Eugenio (1976), *Poet in Our Time*. Marion Boyars Publishers, London, 79 pp. In 1967, Montale was made a lifetime Senator in the Italian Parliament. He received the Nobel Prize for Literature in 1925, the fifth Italian to be so honored.

Montanari, A., Hay, R. L., Alvarez, Walter, Asaro, Frank, Michel, H. V., Alvarez, L. W., and Smit, J. (1983), Spheroids at the Cretaceous–Tertiary boundary are altered impact droplets of basaltic composition. *Geology*, 11, 668-671.

Morton, H. V. (1964), *A Traveller in Italy*. Dodd, Mead and Company, New York, 636 pp. One of the older, standard travel books for Italy, but still very interesting.

Murray, William (1991), *The Last Italian*. Prentice-Hall, New York, 254 pp. William Murray—half-Italian and half-American—writing brilliantly about the real Italy.

Nemerov, Howard (1960), *New and Selected Poems*. The University of Chicago Press, Chicago, 116 pp.

Owen, E. W. (1975), *Trek of the Oil Finders: A History of Exploration for Petroleum*. American Association of Petroleum Geologists, Tulsa, Oklahoma, 1647 pp. Longer than *War and Peace*, but a fine history.

Powell, J. W. (1876), *Report on the Geology of the Eastern Portion of the Uinta Mountains and a Region of Country Adjacent Thereto*. U. S. Geological Survey, 218 pp. A study of 50,000 feet, probably more, of Precambrian, Paleozoic, Mesozoic, and Cenozoic rocks. Work was done in 1868, 1869, 1871, 1874, and 1875. A wonderful book for a geologist—perhaps for anyone—to read. "Some student of geology will eventually find here a subject rich in results," said Powell.

Silone, Ignazio (1933), *Fontamara*. J. M. Dent and Sons, Aldine House, London, 180 pp. Very early exposé of the Fascist horror—before George Orwell and Arthur Koestler. The Abruzzi and its people as only Silone has written.

Silone, Ignazio (1937), *Bread and Wine*. Signet Classic, New American Library, New York, paperback, 286 pp. His best novel.

Smith, W. J., and Gioia, Dana, editors (1985), *Poems from Italy*. New Rivers Press, Bookslinger, St. Paul, Minnesota, 456 pp. Italians erect statues, name streets and piazzas, and throw festivals for poets. Oh that we prized poets as highly as Michael Jordan or even Mike Brown.

Stokes, W. L., Judson, Sheldon, and Picard, M. D. (1978), *Introduction to Geology: Physical and Historical*: Prentice–Hall, Englewood Cliffs, New Jersey, 656 pp. Probably on its way out of print, but one of the few good combined treatments of physical and historical geology.

Thoreau, H. D. (1854), *Walden or Life in the Woods*. Princeton University Press, Princeton, New Jersey, 352 pp. An attractive, inexpensive edition with an introduction by Joyce Carol Oates. One should read *Walden* when young and frequently thereafter.

Updike, John (1985), *Facing Nature*. Alfred A. Knopf, New York, 110 pp. Updike the poet. His novels, short stories, reviews, essays, and poems comprise a colossal body of fine work. Deserves the Nobel Prize for Literature.

About the Author

Born in Missouri, but transported quickly to Wyoming, M. Dane Picard has completed 25 years as Professor of Geology and Geophysics at the University of Utah, Salt Lake City. His 10 years in the oil patch were spent in the Rocky Mountains, but included stints in Los Angeles and Houston. He has been President of the National Association of Geology Teachers, the Society for Sedimentary Geology (SEPM), and the Rocky Mountain section of the American Association of Petroleum Geologists. His more than 150 geology papers have appeared in leading journals here and abroad. He has traveled extensively. His previous books include *Grit and Clay*.